# On Boundary Interpolation
# for Matrix Valued Schur Functions

# MEMOIRS
of the
American Mathematical Society

Number 856

# On Boundary Interpolation for Matrix Valued Schur Functions

Vladimir Bolotnikov
Harry Dym

May 2006 • Volume 181 • Number 856 (end of volume) • ISSN 0065-9266

**American Mathematical Society**
Providence, Rhode Island

2000 *Mathematics Subject Classification.* Primary 30E05, 47A57.

---

**Library of Congress Cataloging-in-Publication Data**

Bolotnikov, Vladimir, 1962–
  On boundary interpolation for matrix valued Schur functions / Vladimir Bolotnikov, Harry Dym.
    p. cm. — (Memoirs of the American Mathematical Society, ISSN 0065-9266 ; no. 856)
  "Volume 181, number 856 (end of volume)."
  Includes bibliographical references.
  ISBN 0-8218-4047-9 (alk. paper)
  1. Schur functions. 2. Interpolation spaces 3. Moment problems (Mathematics) 4. Lyapunov functions. I. Dym, H. (Harry), 1938– II. Title. III. Series.
QA3.A57 no. 856
[QA331.7]
510 s—dc22
[515'.73]                                                                             2006040674

---

## Memoirs of the American Mathematical Society

This journal is devoted entirely to research in pure and applied mathematics.

**Subscription information.** The 2006 subscription begins with volume 179 and consists of six mailings, each containing one or more numbers. Subscription prices for 2006 are US$624 list, US$499 institutional member. A late charge of 10% of the subscription price will be imposed on orders received from nonmembers after January 1 of the subscription year. Subscribers outside the United States and India must pay a postage surcharge of US$31; subscribers in India must pay a postage surcharge of US$43. Expedited delivery to destinations in North America US$35; elsewhere US$130. Each number may be ordered separately; *please specify number* when ordering an individual number. For prices and titles of recently released numbers, see the New Publications sections of the *Notices of the American Mathematical Society*.

**Back number information.** For back issues see the *AMS Catalog of Publications*.

Subscriptions and orders should be addressed to the American Mathematical Society, P. O. Box 845904, Boston, MA 02284-5904, USA. *All orders must be accompanied by payment.* Other correspondence should be addressed to 201 Charles Street, Providence, RI 02904-2294, USA.

**Copying and reprinting.** Individual readers of this publication, and nonprofit libraries acting for them, are permitted to make fair use of the material, such as to copy a chapter for use in teaching or research. Permission is granted to quote brief passages from this publication in reviews, provided the customary acknowledgment of the source is given.

Republication, systematic copying, or multiple reproduction of any material in this publication is permitted only under license from the American Mathematical Society. Requests for such permission should be addressed to the Acquisitions Department, American Mathematical Society, 201 Charles Street, Providence, Rhode Island 02904-2294, USA. Requests can also be made by e-mail to reprint-permission@ams.org.

---

*Memoirs of the American Mathematical Society* is published bimonthly (each volume consisting usually of more than one number) by the American Mathematical Society at 201 Charles Street, Providence, RI 02904-2294, USA. Periodicals postage paid at Providence, RI. Postmaster: Send address changes to Memoirs, American Mathematical Society, 201 Charles Street, Providence, RI 02904-2294, USA.

© 2006 by the American Mathematical Society. All rights reserved.
Copyright of this publication reverts to the public domain 28 years after publication. Contact the AMS for copyright status.
This publication is indexed in *Science Citation Index*®, *SciSearch*®, *Research Alert*®, *CompuMath Citation Index*®, *Current Contents*®/*Physical, Chemical & Earth Sciences*.
Printed in the United States of America.

∞ The paper used in this book is acid-free and falls within the guidelines established to ensure permanence and durability.
Visit the AMS home page at http://www.ams.org/

10 9 8 7 6 5 4 3 2 1        11 10 09 08 07 06

# Contents

1. Introduction — 1
2. Preliminaries — 7
3. Fundamental matrix inequalities — 12
4. On $\mathcal{H}(\Theta)$ spaces — 18
5. Parametrizations of all solutions — 23
6. The equality case — 29
7. Nontangential limits — 32
8. The Nevanlinna–Pick boundary problem — 41
9. A multiple analogue of the Carathéodory–Julia theorem — 48
10. On the solvability of a Stein equation — 65
11. Positive definite solutions of the Stein equation — 76
12. A Carathéodory-Fejér boundary problem — 80
13. The full matrix Carathéodory-Fejér boundary problem — 87
14. The lossless inverse scattering problem — 94

Bibliography — 105

# Abstract

A number of interpolation problems are considered in the Schur class of $p \times q$ matrix valued functions $S$ that are analytic and contractive in the open unit disk. The interpolation constraints are specified in terms of nontangential limits and angular derivatives at one or more (of a finite number of) boundary points. Necessary and sufficient conditions for existence of solutions to these problems and a description of all the solutions when these conditions are met is given. The analysis makes extensive use of a class of reproducing kernel Hilbert spaces $\mathcal{H}(S)$ that was introduced by de Branges and Rovnyak. The Stein equation that is associated with the interpolation problems under consideration is analyzed in detail. A lossless inverse scattering problem is also considered.

---

Received by the editor September 11, 1999 and, in revised form, September 22, 2003.
1991 *Mathematics Subject Classification.* 30E05, 446E22, 47A57.
*Key words and phrases.* Boundary interpolation, Lyapunov-Stein equation, matrix valued Schur functions.
H. Dym thanks Renee and Jay Weiss for endowing the Chair that supports his research and the Israel Science Foundation for support under grant 300/02.

# 1. Introduction

In this paper we shall study a number of tangential interpolation problems in the Schur class of $p \times q$ matrix valued functions that are analytic and contractive in the open unit disk when a finite number of interpolation constraints are imposed on the boundary. We shall work within the framework of the augmented Basic Interpolation Problem (**aBIP**). An introduction to this problem, which includes an account of its development from more elementary problems (such as bitangential versions of the classical Nevanlinna–Pick and Carathéodory–Fejér problems) as well as other formulations, appears in [**27**].

In order to describe the **aBIP** we need to introduce some notation. Let $\mathbf{H}_2^{p \times q}$ denote the set of $\mathbb{C}^{p \times q}$–valued functions with entries in the Hardy space $\mathbf{H}_2$ of the unit disk $\mathbb{D}$ and let $\mathbf{H}_2^{k \times 1}$ be abbreviated by $\mathbf{H}_2^k$. Similarly, let $L_2^k(\mathbb{T})$ designate the Hilbert space of measurable and square integrable $\mathbb{C}^k$–valued functions with inner product

$$\langle f, g \rangle = \frac{1}{2\pi} \int_0^{2\pi} g(e^{it})^* f(e^{it}) dt, \quad (f, g \in L_2^k(\mathbb{T})).$$

The space $\mathbf{H}_2^k$ is identified as the closed subspace of $L_2^k(\mathbb{T})$ which consists of all $f \in L_2^k(\mathbb{T})$ whose Fourier coefficients $f_\ell = \frac{1}{2\pi} \int_0^{2\pi} e^{-i\ell t} f(e^{it}) dt$ are equal to zero for $\ell < 0$. The symbol $\left(\mathbf{H}_2^k\right)^\perp$ stands for the orthogonal complement of $\mathbf{H}_2^k$ with respect to the above inner product. More generally, $\left(\mathbf{H}_2^{p \times q}\right)^\perp$ denote the set of $\mathbb{C}^{p \times q}$–valued functions with entries in $\mathbf{H}_2^\perp$. The Schur class of $\mathbb{C}^{p \times q}$–valued analytic contractions in $\mathbb{D}$ is denoted by $\mathcal{S}^{p \times q}$. In what follows, $\mathbf{H}_\infty^{p \times q}$ will denote the space of $p \times q$ mvf's with entries that are analytic and bounded on $\mathbb{D}$. With every mvf (matrix valued function) $S \in \mathcal{S}^{p \times q}$ we associate the matrix valued Hermitian form $[\,,\,]_S$

$$[h, g]_S = \frac{1}{2\pi} \int_0^{2\pi} g(e^{it})^* \begin{pmatrix} I_p & -S(e^{it}) \\ -S(e^{it})^* & I_q \end{pmatrix} h(e^{it}) dt, \quad (1.1)$$

which is defined for every choice of $h \in L_2^{(p+q) \times k}(\mathbb{T})$ and $g \in L_2^{(p+q) \times \ell}(\mathbb{T})$ and any positive integers $k$ and $\ell$. This form is nonnegative:

$$[h, h]_S \geq 0 \quad \text{for all} \quad h \in L_2^{(p+q) \times k}(\mathbb{T}),$$

since $\begin{pmatrix} I_p & -S(e^{it}) \\ -S(e^{it})^* & I_q \end{pmatrix} \geq 0$ for almost all $t \in [0; 2\pi]$.

Throughout this paper $I_k$ stands for the identity matrix in $\mathbb{C}^{k \times k}$, $J$ denotes the signature matrix defined by

$$J = \begin{pmatrix} I_p & 0 \\ 0 & -I_q \end{pmatrix},$$

and $X^{-*}$ is a convenient shorthand for $(X^*)^{-1}$ when $X$ is invertible. We assume that

$$M, N, P \in \mathbb{C}^{n \times n} \quad \text{and} \quad C \in \mathbb{C}^{(p+q) \times n} \quad (1.2)$$

is a given set of matrices satisfying the Lyapunov–Stein identity

$$M^* P M - N^* P N = C^* J C \quad (1.3)$$

and that the mvf

$$G(z) = M - zN \quad (1.4)$$

is regular:
$$\det G(z) = \det(M - zN) \not\equiv 0. \tag{1.5}$$

The partition
$$C = \begin{pmatrix} C_1 \\ C_2 \end{pmatrix}, \quad C_1 \in \mathbb{C}^{p \times n}, \quad C_2 \in \mathbb{C}^{q \times n}, \tag{1.6}$$

which enables us to express the Lyapunov–Stein equation (1.3) in the form
$$M^*PM - N^*PN = C_1^*C_1 - C_2^*C_2, \tag{1.7}$$

will be useful. The symbol $\widehat{\mathbf{aBIP}}(M, N, P, C)$ will be used to denote the following interpolation problem (which is a relaxed version of the $\mathbf{aBIP}$ that will be discussed below):

(1) *Find necessary and sufficient conditions which insure the existence of a Schur function $S \in \mathcal{S}^{p \times q}$ such that*
$$\begin{pmatrix} I_p & -S(\zeta) \\ -S(\zeta)^* & I_q \end{pmatrix} CG(\zeta)^{-1} \in \begin{pmatrix} \mathbf{H}_2^{p \times n} \\ (\mathbf{H}_2^{q \times n})^\perp \end{pmatrix} \tag{1.8}$$

and
$$P_S \leq P, \tag{1.9}$$

where
$$P_S := \frac{1}{2\pi} \int_0^{2\pi} G(e^{it})^{-*}C^* \begin{pmatrix} I_p & -S(e^{it}) \\ -S(e^{it})^* & I_q \end{pmatrix} CG(e^{it})^{-1} dt$$
$$= [CG(\zeta)^{-1}, CG(\zeta)^{-1}]_S. \tag{1.10}$$

(2) *Describe the set $\widehat{\mathcal{S}}(M, N, P, C)$ of all such mvf's.*

If the mvf $G(z)$ is not invertible everywhere on $\mathbb{T}$, the condition (1.9) is meant in the following sense: the integral
$$\frac{1}{2\pi} \int_0^{2\pi} G(e^{it})^{-*}C^* \begin{pmatrix} I_p & -S(e^{it}) \\ -S(e^{it})^* & I_q \end{pmatrix} CG(e^{it})^{-1} dt \tag{1.11}$$

converges to a matrix $P_S$ which satisfies inequality (1.9).

The $\widehat{\mathbf{aBIP}}(M, N, P, C)$ can be formulated without mention of the space $\mathbf{H}_2^\perp$ with the help of the symbol
$$H(z) = zM^* - N^* \tag{1.12}$$

and the following remark:

REMARK 1.1. Condition (1.8) is equivalent to the following two conditions
$$B(\zeta) := \begin{pmatrix} I_p, & -S(\zeta) \end{pmatrix} CG(\zeta)^{-1} \in \mathbf{H}_2^{p \times n} \tag{1.13}$$

and
$$\widetilde{B}(\zeta) := H(\zeta)^{-1} C^* \begin{pmatrix} -S(\zeta) \\ I_q \end{pmatrix} \in \mathbf{H}_2^{n \times q}. \tag{1.14}$$

PROOF. It is easily seen that a function $f(\zeta)$ belongs to $\mathbf{H}_2$ if and only if the function $\bar{\zeta} f(\zeta)^*$ belongs to $\mathbf{H}_2^\perp$. Therefore,
$$H(\zeta)^{-1}C^* \begin{pmatrix} -S(\zeta) \\ I_q \end{pmatrix} \in \mathbf{H}_2^{n \times q} \iff \bar{\zeta} \begin{pmatrix} -S(\zeta)^*, & I_q \end{pmatrix} CH(\zeta)^{-*} \in (\mathbf{H}_2^{q \times n})^\perp$$

and, since
$$\bar{\zeta} H(\zeta)^{-*} = G(\zeta)^{-1} \quad \text{for } \zeta \in \mathbb{T}, \tag{1.15}$$
condition (1.14) is equivalent to
$$\begin{pmatrix} -S(\zeta)^*, & I_q \end{pmatrix} CG(\zeta)^{-1} \in (\mathbf{H}_2^{q \times n})^\perp,$$
which together with (1.13) is equivalent to (1.8). □

REMARK 1.2. Since the form $[\,,\,]_S$ is nonnegative, it follows that the matrix $P_S$ defined by (1.10) is positive semidefinite. Moreover, it satisfies the Lyapunov–Stein equation
$$M^* P_S M - N^* P_S N = C^* JC \tag{1.16}$$
and is subject to
$$\bigcap_{\substack{\zeta \in \mathbb{T} \\ \det G(\zeta) \neq 0}} \operatorname{Ker} CG(\zeta)^{-1} \subseteq \operatorname{Ker} P_S. \tag{1.17}$$

For a proof of (1.16), see e.g., [15, Lemma 2.1]; the inclusion (1.17) follows readily from (1.10) and means that every vector $f \in \mathbb{C}^n$ such that $CG(z)^{-1} f \equiv 0$, belongs to $\operatorname{Ker} P_S$ (if the columns of $CG(z)^{-1}$ are linearly independent, then the identity $CG(z)^{-1} f \equiv 0$ forces $f = 0$, so the conclusion $f \in \operatorname{Ker} P_S$ does not convey any new information).

It follows from (1.9) that the condition
$$P \geq 0 \tag{1.18}$$
is necessary for the $\widehat{\mathbf{aBIP}}(M, N, P, C)$ to have a solution. It turns out that this condition is also sufficient. Moreover, if $P \geq 0$, then the set $\widehat{\mathcal{S}}(M, N, P, C)$ can be parametrized in terms of a linear fractional transformation; see Theorem 5.1 below.

Theorem 5.1 was proved in [15] by adapting Potapov's method of Fundamental Matrix Inequalities (FMI's) to the $\widehat{\mathbf{aBIP}}$ framework. Additional analysis of the FMI corresponding to the $\widehat{\mathbf{aBIP}}$ is given in Section 3. There we also discuss connections between Potapov's method and the reproducing kernel Hilbert space methods which are frequently used for the investigation of interpolation problems. Some preliminary facts about positive kernels and some concrete reproducing kernel Hilbert spaces that are needed in the sequel are given in Sections 2 and 4.

The $\widehat{\mathbf{aBIP}}(M, N, P, C)$ is termed *nondegenerate* if $P > 0$ and is termed *degenerate* if $P$ is singular (and positive semidefinite). In Section 5 we present an alternative parametrization of the set $\widehat{\mathcal{S}}(M, N, P, C)$ in terms of a Redheffer linear fractional transformation (see Theorem 5.9 below) which sometimes (especially in the degenerate case) turns to be more convenient for applications (see, e.g., Sections 11 and 12 in [15]). This approach has been adapted in [27] and [15] from the work of Katsnelson, Kheifets and Yuditskii [36] on the abstract interpolation problem and the work of Arov and Grossman [9], [10] on the coupling of unitary colligations.

The problem of describing all functions $S \in \mathcal{S}^{p \times q}$ for which equality prevails in (1.9) is quite different. For such a problem the matrix $P$ has to inherit the properties (1.16) and (1.17) of $P_S$. More precisely, assuming that (1.5) is in force and that $P$ is a positive semidefinite solution of (1.7), let $\mathbf{aBIP}(M, N, P, C)$ denote the following "augmented" basic interpolation problem:

(1) *Find necessary and sufficient conditions which insure the existence of a Schur function $S \in \mathcal{S}^{p \times q}$ satisfying (1.8) and the equality*

$$P_S := \left[CG(\zeta)^{-1}, \, CG(\zeta)^{-1}\right]_S = P. \tag{1.19}$$

(2) *Describe the set $\mathcal{S}(M, N, P, C)$ of all such functions.*

It is easily seen that

$$\mathcal{S}(M, N, P, C) \subseteq \widehat{\mathcal{S}}(M, N, P, C). \tag{1.20}$$

By the preceding discussion, the conditions (1.3), (1.18) and

$$\bigcap_{\substack{\zeta \in \mathbb{T} \\ \det G(\zeta) \neq 0}} \operatorname{Ker} CG(\zeta)^{-1} \subseteq \operatorname{Ker} P \tag{1.21}$$

are necessary for the $\mathbf{aBIP}(M, N, P, C)$ to have a solution. However, they are not sufficient. A criterion for the solvability of the $\mathbf{aBIP}(M, N, P, C)$ will be given in Section 6 and adapted to two concrete cases in Sections 8 and 12. In the former we also present a matrix analogue of conditions for the solvability of a boundary Nevanlinna–Pick problem that was obtained in a recent paper of Sarason [**50**]. Another approach to handling boundary interpolation problems in the scalar case is presented in [**29**].

We remark that

$$\bigcap_{\substack{\zeta \in \mathbb{T} \\ \det G(\zeta) \neq 0}} \operatorname{Ker} CG(\zeta)^{-1} = \bigcap_{j=0}^{n-1} \operatorname{Ker} C\left(G(\alpha)^{-1}N\right)^j$$

for any point $\alpha \in \mathbb{C}$ at which $G(\alpha)$ is invertible. Thus, in the special case when $P > 0$ and $M = I_n$, condition (1.21) holds if and only if

$$\bigcap_{j=0}^{n-1} \operatorname{Ker} CN^j = \{0\},$$

i.e., if and only if the pair $(C, N)$ is observable. There is an analogous interpretation in the more general setting (when $P$ is still positive definite but $M$ is not necessarily invertible); the relevant theory in this general setting is developed in extensive detail in [**6**].

In [**15**] the $\mathbf{aBIP}(M, N, P, C)$ was considered under the assumption that $G(z)$ is invertible at every point on the unit circle:

$$\det G(\zeta) = \det (M - \zeta N) \neq 0 \quad \text{for} \quad |\zeta| = 1, \tag{1.22}$$

which of course, is a more restrictive condition than (1.5). It was shown in [**15**, Lemma 3.1] that under assumption (1.22) the problems $\mathbf{aBIP}(M, N, P, C)$ and $\widehat{\mathbf{aBIP}}(M, N, P, C)$ are equivalent, i.e., the sets $\mathcal{S}(M, N, P, C)$ and $\widehat{\mathcal{S}}(M, N, P, C)$ of their solutions coincide. For this case the interpolation conditions of the $\mathbf{aBIP}$ can be expressed in terms of contour integrals. The use of contour integrals to formulate interpolation conditions was suggested by A. Nudelman in [**43**] and was utilized in [**13**] for the case when the associated Pick matrix $P$ is invertible; for additional discussion and comparison, see [**25**, Section 7.4].

In this paper we shall focus on the opposite "extreme": the case when all the singular points of $G^{-1}$ fall on $\mathbb{T}$:
$$\det(M - zN) \neq 0 \quad \text{if } z \notin \mathbb{T}. \tag{1.23}$$
In this case (1.8) is an automatic consequence of (1.9):

LEMMA 1.3. *Let (1.23) be in force, and let $S \in \mathcal{S}^{p \times q}$ satisfy (1.9) for some positive semidefinite matrix $P \in \mathbb{C}^{n \times n}$. Then $S$ also satisfies condition (1.8) (or equivalently, conditions (1.13) and (1.14)).*

PROOF. To show that the mvf's $B$ and $\widetilde{B}$ defined in (1.13) and (1.14) belong to $\mathbf{H}_2^{p \times n}$ and $\mathbf{H}_2^{n \times q}$, respectively, we shall use the maximum principle for Smirnov class functions. We recall that a mvf which is analytic in $\mathbb{D}$ is said to belong to the *Smirnov class* $\mathcal{N}_+^{p \times q}$ if it can be represented as the ratio of a $\mathbf{H}_\infty^{p \times q}$-function and a scalar $\mathbf{H}_\infty(\mathbb{D})$-function which is outer. For mvf's in this Smirnov class the following maximum principle holds:
$$\mathcal{N}_+^{p \times q} \cap L_2^{p \times q}(\mathbb{T}) = \mathbf{H}_2^{p \times q};$$
for more information on matrix valued Smirnov classes see [**8**], [**37**].

Let $B$ be defined by (1.13). By (1.23), all the roots of the scalar polynomial $\det G(z)$ fall on $\mathbb{T}$ and hence, $\det G(z)$ is outer in $\mathbf{H}_\infty(\mathbb{D})$. The rational mvf $G^{-1}$ admits a representation of the form
$$G(z)^{-1} = \frac{G_1(z)}{\det G(z)},$$
where $G_1$ is a $\mathbb{C}^{n \times n}$-valued polynomial. Thus the mvf
$$B(z) = \frac{(I_p, -S(z))\, CG_1(z)}{\det G(z)}$$
belongs to the matrix Smirnov class $\mathcal{N}_+^{p \times n}$, since $(I_p, -S(z))\, CG_1(z)$ belongs to $\mathbf{H}_\infty^{p \times n}$.

Making use of decomposition (1.6) and definition (1.10), we rewrite (1.9) as
$$\frac{1}{2\pi} \int_0^{2\pi} G(e^{it})^{-*} C^* \begin{pmatrix} I_p \\ -S(e^{it})^* \end{pmatrix} (I_p, -S(e^{it}))\, CG(e^{it})^{-1} dt$$
$$\leq P - \frac{1}{2\pi} \int_0^{2\pi} G(e^{it})^{-*} C_2^* \left(I_q - S(e^{it})^* S(e^{it})\right) C_2 G(e^{it})^{-1} dt$$
and conclude that $B(e^{it})$ belongs to $L_2^{p \times n}$. By the maximum principle for Smirnov class functions, $B \in \mathbf{H}_2^{p \times n}$.

Next, since the scalar polynomial $\det H(z)$ is outer in $\mathbf{H}_\infty(\mathbb{D})$, it follows much as above that the function $\widetilde{B}$ defined in (1.14) belongs to the Smirnov class $\mathcal{N}_+^{n \times q}$. Taking advantage of (1.15), we rewrite (1.9) as
$$\frac{1}{2\pi} \int_0^{2\pi} H(e^{it})^{-1} C^* \begin{pmatrix} -S(e^{it}) \\ I_q \end{pmatrix} (-S(e^{it})^*, I_q)\, CH(e^{it})^{-*} dt$$
$$\leq P - \frac{1}{2\pi} \int_0^{2\pi} G(e^{it})^{-*} C_1^* \left(I_p - S(e^{it}) S(e^{it})^*\right) C_1 G(e^{it})^{-1} dt.$$

Therefore, $\widetilde{B}(e^{it})$ belongs to $L_2^{n\times q}$ and, by the Smirnov maximum principle, $\widetilde{B} \in \mathbf{H}_2^{n\times q}$, which completes the proof. □

COROLLARY 1.4. *Let* (1.23) *be in force, let* $P$ *be a positive semidefinite solution of the Lyapunov–Stein equation* (1.7) *and let* $S \in \mathcal{S}^{p\times q}$ *satisfy* (1.9). *Then* $S$ *belongs to* $\widehat{\mathcal{S}}(I_n, N, P, C)$.

Under condition (1.23), we may assume without loss of generality (see Remark 2.12 below) that $M = I_n$ and $\mathrm{spec}(N) \subseteq \mathbb{T}$. In this case every positive semidefinite solution of the corresponding Stein equation

$$P - N^*PN = C_1^*C_1 - C_2^*C_2 \qquad (1.24)$$

is partly specified by $N$, $C_1$ and $C_2$. Two special choices of $N$ are considered in detail:

$$N = \begin{pmatrix} \bar{\beta}_1 I_{r_1} & & \\ & \ddots & \\ & & \bar{\beta}_m I_{r_m} \end{pmatrix} \quad \text{and} \quad N = \begin{pmatrix} \bar{\beta} I_r & I_r & & \\ & \bar{\beta} I_r & \ddots & \\ & & \ddots & I_r \\ & & & \bar{\beta} I_r \end{pmatrix}, \qquad (1.25)$$

where $\beta_1, \ldots, \beta_m$ and $\beta$ are points on $\mathbb{T}$.

It will be shown in Section 8 that for the first choice of $N$, the off diagonal blocks of every solution $P$ of the Stein equation (1.24) are uniquely specified, whereas the diagonal blocks of $P$ may be chosen freely. By taking advantage of this freedom, it is easily seen that, for this choice of $N$, the Stein equation (1.24) always has a positive semidefinite (and even a positive definite) solution $P$. It will be shown in Section 8 that the corresponding $\widehat{\mathbf{aBIP}}(I_n, N, P, C)$ is equivalent to a Nevanlinna–Pick boundary problem which is usually posed in terms of nontangential boundary limits. By including this problem in the general framework of the $\widehat{\mathbf{aBIP}}$, we are able to describe the set of all its solutions using general results from Section 5.

It turns out that for the second choice of $N$ in (1.25), the Stein equation (1.24) may not have a positive semidefinite solution. The structure of the set of all solutions $P$ of (1.24) and necessary and sufficient conditions for (1.24) to have a positive definite (or positive semidefinite solutions) are discussed in Sections 10 and 11. The corresponding $\widehat{\mathbf{aBIP}}(I_n, N, P, C)$ turns to be equivalent to a certain tangential Carathéodory-Fejér boundary problem. The block entries of $P$ which are not specified by (1.24), are related to certain nontangential boundary limits of the interpolant $S$. This is clarified in Section 12. To this end, we use some auxiliary results on the boundary behavior of analytic functions in $\mathbb{D}$ which are collected in Section 7.

In Section 13 we consider another Carathéodory-Fejér boundary problem (the full matrix interpolation problem) which was studied by Kovalishina in [**40**] and show how to incorporate this problem into the general scheme of the $\widehat{\mathbf{aBIP}}$. This identification enables us to obtain a description of the set of all solutions even when the Pick matrix $P$ is singular. A number of related results obtained by Kovalishina are also discussed in some detail.

The classical theorems of Carathéodory and Julia on angular derivatives (at the boundary) of Schur functions may be found in [**51**, Chapter 4] and [**49**, Chapter

6]. For additional sources and discussion see the Notes at the end of each of these chapters. Generalizations to matrix valued Schur functions were considered in [**39**]; for tangential versions, see [**23**, pp. 97–99] and Section 8 below. In Section 9 we present a tangential analogue for higher order angular derivatives for matrix valued Schur functions.

The analysis in both Sections 8 and 9 (as well as in a number of other sections) makes extensive use of the reproducing kernel Hilbert spaces $\mathcal{H}(S)$ for $p \times q$ matrix valued Schur functions (see Section 2 below for the precise definition) that were introduced by de Branges and Rovnyak in [**17**] and [**18**]. The use of these spaces to study angular derivatives seems to have been initiated in [**23**, Chapter 8] and, independently for scalar Schur functions, by Sarason in [**48**]; see also his monograph [**49**] for further extensions. Some general classes of interpolation problems for square mvf's that include constraints on the boundary have also been considered from another point of view in [**12**] and [**14**].

The results of Sections 9 and 12 are used in Section 14, where the lossless inverse scattering problem (**LISP**) is discussed. This problem may be stated as follows:
*Given* $S \in \mathcal{S}^{p \times q}$, *find all* $J$–*inner mvf's*[1] $\Theta$ *that are analytic in* $\mathbb{D}$ *and meet the constraint*

$$(I_p, -S(z))\Theta(z)J\Theta(z)^* \begin{pmatrix} I_p \\ -S(z) \end{pmatrix} \geq 0 \qquad (|z| < 1).$$

The **LISP** was first considered for scalar Schur functions in [**20**]. Therein necessary and sufficient conditions for the existence of a rational solution with one or more poles on the boundary were expressed in terms of the representing measure for the Carathéodory function[2] $(1+S)(1-S)^{-1}$. This study was motivated by questions in network theory and stochastic estimation theory. Some other applications of boundary interpolation problems are indicated in [**30**].

The general **LISP** for square matrix valued Schur functions was solved in [**3**]. There too it proved convenient to work with the (now matrix valued) Carathéodory function $(I_p + S)(I_p - S)^{-1}$. The **LISP** for general $p \times q$ matrix valued Schur functions was considered in [**23**, Section 8]. An explicit construction of all rational solutions which are analytic in the closed unit disk $\overline{\mathbb{D}}$ and of the solutions with one simple pole on the boundary was given there. In Section 14 we shall extend this analysis to obtain a description of all the rational solutions of the **LISP** with an arbitrary number of poles on the boundary, simple or not. There too conditions for the existence of a solution to this problem will be formulated directly in terms of $S$.

## 2. Preliminaries

A $n \times n$ mvf $K_\omega(z)$ defined on $\Omega \times \Omega$ is said to be *a positive kernel* if

$$\sum_{i,j=1}^{r} u_i^* K_{\omega_j}(\omega_i) u_j \geq 0$$

---

[1]The precise definition of a $J$–inner mvf will be given in Section 4.
[2]See Section 3 for the definition.

for every choice of an integer $r \geq 1$, of points $\omega_1, \ldots, \omega_r \in \Omega$ and of vectors $u_1, \ldots, u_r \in \mathbb{C}^n$, or, equivalently, if the Hermitian block matrix with $ij$-th entry $K_{\omega_j}(\omega_i)$ is positive semidefinite. This property of $K$ will be denoted by $K_\omega(z) \succeq 0$ and we write $K_\omega^1(z) \succeq K_\omega^2(z)$ if $K_\omega^1(z) - K_\omega^2(z) \succeq 0$. It is readily checked that if $K_\omega(z) \succeq 0$, then $K_\omega(z)^* = K_z(\omega)$. The following simple observations follow readily from the definition of a positive kernel and will be useful.

PROPOSITION 2.1. *Let $K_\omega(z)$ be a positive kernel on $\Omega \times \Omega$ and let $T_1(z)$ and $T_2(z)$ be two mvf's of appropriate sizes. Then*

$$\begin{aligned}(T_1(z) + T_2(z)) K_\omega(z) (T_1(\omega)^* + T_2(\omega)^*) &\preceq 2T_1(z) K_\omega(z) T_1(\omega)^* \\ &\quad + 2T_2(z) K_\omega(z) T_2(\omega)^*.\end{aligned}$$

PROPOSITION 2.2. *Let $K_\omega^1(z) \succeq K_\omega^2(z) \succeq 0$ on $\Omega \times \Omega$ and let $\|K_\omega^1(z)\| \leq \gamma$ for every pair of points $z, \omega \in \Omega$. Then*

$$\|K_\omega^2(z)\| \leq \gamma \quad (\forall\, z, \omega \in \Omega). \tag{2.1}$$

PROOF. Since $K_\omega^2(z) \succeq 0$, it follows that for every $x, y \in \mathbb{C}^n$,

$$|x^* K_\omega^2(z) y|^2 \leq (x^* K_z^2(z) x)(y^* K_\omega^2(\omega) y) \leq (x^* K_z^1(z) x)(y^* K_\omega^1(\omega) y) \leq \gamma^2 \|x\|^2 \|y\|^2,$$

which leads easily to (2.1). □

If $K_\omega(z)$ is continuous on $\Omega \times \Omega$, then

$$K_\omega(z) \succeq 0 \quad \text{if and only if} \quad \int_\Gamma \int_\Gamma \varphi(\varsigma)^* K_\varsigma(\zeta) \varphi(\zeta) \, d\zeta d\bar\varsigma \geq 0 \tag{2.2}$$

for every simple curve $\Gamma \subset \Omega$ and every $\mathbb{C}^n$-valued function $\varphi$ which is continuous on $\Gamma$. This equivalence was established for open subintervals $\Omega$ of $\mathbb{R}$ in [34] and for arbitrary open subsets $\Omega$ of $\mathbb{R}$ in [45, Section 2.12]. Therefore, since the same arguments are applicable for open subsets $\Omega$ of $\mathbb{C}$, we are led to the following result:

PROPOSITION 2.3. *Let $\Omega \subset \mathbb{C}$ be a simply connected open set and let the kernel $K_\omega(z)$ be positive on $\Omega \times \Omega$ and analytic in $z$ and $\bar\omega$. Furthermore, let $A(z)$ be analytic in $\Omega$. Then for every nonnegative integer $m$, the kernel*

$$K_\omega^{(m)}(z) := \frac{\partial^{2m}}{\partial z^m \partial \bar\omega^m} (A(z) K_\omega(z) A(\omega)^*)$$

*is also positive on $\Omega \times \Omega$.*

PROOF. Take an integer $r$ and points $\omega_1, \ldots, \omega_r \in \Omega$ and let $\Gamma$ be a closed simple contour in $\Omega$ which surrounds all of these points. Then

$$\begin{aligned}\left(K_{\omega_i}^{(m)}(\omega_j)\right)_{i,j=1}^r &= \left(\frac{\partial^{2m}}{\partial z^m \partial \bar\omega^m} (A(z) K_\omega(z) A(\omega)^*)|_{z=\omega_j,\, \omega=\omega_i}\right)_{i,j=1}^r \\ &= \left(\frac{m!}{2\pi}\right)^2 \int_\Gamma \int_\Gamma \varphi(\varsigma)^* A(\zeta) K_\varsigma(\zeta) A(\varsigma)^* \varphi(\zeta) \, d\zeta d\bar\varsigma,\end{aligned}$$

where

$$\varphi(\zeta) = \left(\frac{I_n}{(\zeta - \omega_1)^{m+1}}, \ldots, \frac{I_n}{(\zeta - \omega_r)^{m+1}}\right).$$

By (2.2), the matrix on the left hand side of the latter chain of equalities is positive semidefinite. Therefore, $K_\omega^{(m)}(z) \succeq 0$. □

In the same spirit one can prove the following slightly more general statement.

PROPOSITION 2.4. *Let $\Omega \subset \mathbb{C}$ be a simply connected open set, let the kernel $K_\omega(z)$ be positive on $\Omega \times \Omega$ and analytic in $z$ and $\bar{\omega}$ and let $A_1(z), \ldots, A_k(z)$ be mvf's which are analytic in $\Omega$. Then for every choice of nonnegative integers $m_1, \ldots, m_k$ the block matrix kernel $\mathbf{K}(z, \omega)$ with the block entries*

$$\mathbf{K}_{j\ell}(z,\omega) := \frac{\partial^{m_j+m_\ell}}{\partial z^{m_j} \partial \bar{\omega}^{m_\ell}} \left( A_j(z) K_\omega(z) A_\ell(\omega)^* \right) \quad (j,\ell = 1,\ldots,k)$$

*is positive on $\Omega \times \Omega$.*

A Hilbert space $\mathcal{H}(K)$ of $\mathbb{C}^n$-valued functions which are defined on a subset $\Omega$ of $\mathbb{C}$ is said to be a reproducing kernel Hilbert space with reproducing kernel $K_\omega(z)$ if for every point $\omega \in \Omega$ and every vector $x \in \mathbb{C}^n$, the function $K_\omega(z)x$ belongs to $\mathcal{H}(K)$ as a function of $z$ and

$$\langle f, K_\omega x \rangle_{\mathcal{H}(K)} = x^* f(\omega) \qquad (\forall\, f \in \mathcal{H}(K)). \tag{2.3}$$

Formula (2.3) implies that

$$\sum_{i,j=1}^r x_i^* K_{\omega_j}(\omega_i) x_j = \left\langle \sum_{j=1}^r K_{\omega_j} x_j, \sum_{i=1}^r K_{\omega_i} x_i \right\rangle_{\mathcal{H}(K)} \geq 0 \qquad (x_j \in \mathbb{C}^n,\ \omega_j \in \Omega)$$

and hence exhibits the reproducing kernel $K_\omega(z)$ of $\mathcal{H}(K)$ as a positive kernel. Conversely, by the matrix version of a theorem of Aronszajn [**7**], every positive kernel $K_\omega(z)$ on $\Omega \times \Omega$ can be identified as the reproducing kernel of such a reproducing kernel Hilbert space.

The following proposition characterizes $\mathcal{H}(K)$ in terms of positive kernels (for a proof, see [**47**] for the scalar case and [**1**, Lemma 2.2] for the matrix case).

PROPOSITION 2.5. *A vector valued function $f$ defined on $\Omega$ belongs to $\mathcal{H}(K)$ and satisfies $\|f\|^2_{\mathcal{H}(K)} \leq \gamma$ if and only if the following kernel is positive on $\Omega \times \Omega$:*

$$K_\omega(z) - \gamma^{-1} f(z) f(\omega)^* \succeq 0.$$

EXAMPLE 2.6. The Hardy space $\mathbf{H}_2^n$ is a reproducing kernel Hilbert space with the reproducing kernel $K_\omega(z) = \dfrac{I_n}{\rho_\omega(z)}$, where

$$\rho_\omega(z) = 1 - z\bar{\omega}. \tag{2.4}$$

In this case, formula (2.3) is just Cauchy's formula for $\mathbf{H}_2^n$.

EXAMPLE 2.7. Let $P \in \mathbb{C}^{n \times n}$ be a positive definite matrix with entries $P_{ij}$, let $f_1, \ldots, f_n$ be a set of $n$ linearly independent $k \times 1$ vector functions which are meromorphic in $\mathbb{D}$ and let $\mathcal{H}$ be their span endowed with the inner product based on

$$\langle f_i, f_j \rangle_\mathcal{H} = P_{ij}.$$

Then $\mathcal{H}$ is a reproducing kernel Hilbert space with reproducing kernel

$$K_\omega(z) = F(z) P^{-1} F(\omega)^*, \quad \text{where} \quad F(z) = (f_1(z),\ \ldots,\ f_n(z)).$$

If $S \in \mathcal{S}^{p \times q}$, then
$$\Lambda_\omega(z) = \frac{I_q - S(z)S(\omega)^*}{\rho_\omega(z)} \tag{2.5}$$
is a positive kernel on $\mathbb{D} \times \mathbb{D}$. The corresponding reproducing kernel Hilbert space will be referred as to $\mathcal{H}(S)$.

The following alternate characterization of $\mathcal{H}(S)$, as the space of all vector functions $f \in \mathbf{H}_2^p$ such that
$$\kappa(f) := \sup_{g \in \mathbf{H}_2^q} \left\{ \|f + Sg\|_{\mathbf{H}_2^p}^2 - \|g\|_{\mathbf{H}_2^q}^2 \right\} \tag{2.6}$$
is finite and $\|f\|_{\mathcal{H}(S)}^2 = \kappa(f)$, is due to de Branges and Rovnyak [17], [18].

The next lemma expresses the reproducing kernel $\Lambda_\omega$ of the space $\mathcal{H}(S)$ in terms of the nonnegative form $[\,,\,]_S$ defined via (1.1).

LEMMA 2.8. *The formula*
$$\Lambda_\omega(z) = \left[ \begin{pmatrix} I_p \\ S(\omega)^* \end{pmatrix} \frac{1}{\rho_\omega}, \begin{pmatrix} I_p \\ S(z)^* \end{pmatrix} \frac{1}{\rho_z} \right]_S \tag{2.7}$$
*is valid for every pair of points $z$ and $\omega$ in $\mathbb{D}$.*

PROOF. It is readily seen that the functions
$$f_z(\zeta) = \begin{pmatrix} I_p \\ S(z)^* \end{pmatrix} \frac{1}{\rho_z(\zeta)} \quad \text{and} \quad g_\omega(\zeta) = \frac{S(\omega)^* - S(\zeta)^*}{\rho_\omega(\zeta)} \tag{2.8}$$
belong to $\mathbf{H}_2^{(p+q) \times p}$ and $(\mathbf{H}_2^{q \times p})^\perp$, respectively. Thus, as $\dfrac{I_p}{\rho_\omega(z)}$ is the reproducing kernel for $\mathbf{H}_2^p$, it follows from definition (1.1) that
$$\begin{aligned}[][f_\omega y, f_z x]_S &= \left\langle \begin{pmatrix} I_p & -S \\ -S^* & I_q \end{pmatrix} \begin{pmatrix} I_p \\ S(\omega)^* \end{pmatrix} \frac{y}{\rho_\omega}, \begin{pmatrix} I_p \\ S(z)^* \end{pmatrix} \frac{x}{\rho_z} \right\rangle_{L_2^{p+q}(\mathbb{T})} \\ &= \left\langle \Lambda_\omega y, \frac{x}{\rho_z} \right\rangle_{\mathbf{H}_2^p} + \left\langle g_\omega y, \frac{S(z)^* x}{\rho_z} \right\rangle_{L_2^q(\mathbb{T})} = x^* \Lambda_\omega(z) y, \end{aligned}$$
for every choice of $x$ and $y$ in $\mathbb{C}^p$ and hence that (2.7) is valid. □

The following simple observation will be useful.

LEMMA 2.9. *Let $[\,,\,]_S$ be the form defined in (1.1) and let $F_\omega(\zeta)$ be a $(p+q) \times n$ mvf defined for $\zeta$ and $\omega$ in $\mathbb{D}$. Then the kernel*
$$\mathbf{F}_\omega(z) := [F_\omega, F_z]_S$$
*is positive on $\mathbb{D}$.*

PROOF. For every choice of an integer $r$ and of points $\omega_1, \ldots, \omega_r \in \mathbb{D}$,
$$\begin{aligned} \left( \mathbf{F}_{\omega_j}(\omega_i) \right)_{i,j=1}^r &= \left( [F_{\omega_j}, F_{\omega_i}]_S \right)_{i,j=1}^r \\ &= \frac{1}{2\pi} \int_0^{2\pi} \widehat{F}(e^{it})^* \begin{pmatrix} I_p & -S(e^{it}) \\ -S(e^{it})^* & I_q \end{pmatrix} \widehat{F}(e^{it}) dt, \end{aligned}$$
where
$$\widehat{F}(\zeta) = (F_{\omega_1}(\zeta), \ldots, F_{\omega_r}(\zeta)).$$
Thus, the matrix on the left hand side of the second line of the proof is positive semidefinite. Therefore, $\mathbf{F}_\omega(z) \succeq 0$. □

## 2. PRELIMINARIES

We conclude this section with the following two lemmas which allow us to express the mvf $G(z)$ defined by (1.4) in a certain canonical form which will simplify some later computations.

LEMMA 2.10. *Let $T_1$ and $T_2$ be two invertible $n \times n$ matrices and let*
$$\widehat{M} = T_1 M T_2, \quad \widehat{N} = T_1 N T_2, \quad \widehat{P} = T_1^{-*} P T_1^{-1}, \quad \text{and} \quad \widehat{C} = C T_2. \qquad (2.9)$$
*Then the problems* **aBIP**$(M, N, P, C)$ *and* $\widehat{\textbf{aBIP}}(M, N, P, C)$ *are equivalent to* **aBIP**$(\widehat{M}, \widehat{N}, \widehat{P}, \widehat{C})$ *and* $\widehat{\textbf{aBIP}}(\widehat{M}, \widehat{N}, \widehat{P}, \widehat{C})$, *respectively.*

PROOF. It suffices to note that
$$\widehat{M}^* \widehat{P} \widehat{M} - \widehat{N}^* \widehat{P} \widehat{N} = \widehat{C}^* J \widehat{C},$$
and that
$$\widehat{C}\widehat{G}(z)^{-1} := \begin{pmatrix} \widehat{C}_1 \\ \widehat{C}_2 \end{pmatrix} (\widehat{M} - z\widehat{N})^{-1}$$
$$= \begin{pmatrix} C_1 \\ C_2 \end{pmatrix} (M - zN)^{-1} T_1^{-1} = CG(z)^{-1} T_1^{-1},$$
which implies that
$$\widehat{P}_S = \left[\widehat{C}\widehat{G}(\zeta)^{-1}, \widehat{C}\widehat{G}(\zeta)^{-1}\right]_S = \left[CG(\zeta)^{-1} T_1^{-1}, CG(\zeta)^{-1} T_1^{-1}\right]_S$$
$$= T_1^{-*} \left[CG(\zeta)^{-1}, CG(\zeta)^{-1}\right]_S T_1^{-1}$$
$$= T_1^{-*} P_S T_1^{-1}.$$

□

The next lemma (for the proof see [**15**, Lemma 2.3]) is a slight variation of the canonical representation for regular pencils (see e.g., [**28**, p. 28, Theorem 3]).

LEMMA 2.11. *Let $M$ and $N$ satisfy (1.5). Then there exist invertible matrices $T_1$ and $T_2$ from $\mathbb{C}^{n \times n}$ and matrices $A_1 \in \mathbb{C}^{k_1 \times k_1}$, $A_2 \in \mathbb{C}^{k_2 \times k_2}$ and $A_3 \in \mathbb{C}^{k_3 \times k_3}$ with*
$$\text{spec } A_1 \bigcup \text{spec } A_2 \subset \mathbb{D} \quad \text{and} \quad \text{spec } A_3 \subset \mathbb{T} \qquad (2.10)$$
*such that*
$$T_1 M T_2 = \begin{pmatrix} I_{k_1} & 0 & 0 \\ 0 & A_2 & 0 \\ 0 & 0 & I_{k_3} \end{pmatrix} \quad \text{and} \quad T_1 N T_2 = \begin{pmatrix} A_1 & 0 & 0 \\ 0 & I_{k_2} & 0 \\ 0 & 0 & A_3 \end{pmatrix}.$$

The following remark is an immediate consequence of the last two lemmas.

REMARK 2.12. *Let condition (1.5) be in force. Then, without loss of generality, the matrices $M$ and $N$ from the data set (1.2) of the $\widehat{\textbf{aBIP}}$ can be assumed to be of the form*
$$M = \begin{pmatrix} I_{k_1} & 0 & 0 \\ 0 & A_2 & 0 \\ 0 & 0 & I_{k_3} \end{pmatrix} \quad \text{and} \quad N = \begin{pmatrix} A_1 & 0 & 0 \\ 0 & I_{k_2} & 0 \\ 0 & 0 & A_3 \end{pmatrix}, \qquad (2.11)$$
*where the $A_j$ are matrices satisfying (2.10) and can be presumed to be in Jordan form.*

## 3. Fundamental matrix inequalities

In [**15**] the $\widehat{\mathbf{aBIP}}$ was considered using two different approaches. One of them was based on Potapov's method (which characterizes the solutions of an interpolation problem in terms of a related fundamental matrix inequality; see e.g. [**41**]) suitably adapted to the present framework. The next theorem extends that analysis (especially [**15**, Lemma 3.5]) and characterizes all the solutions of the $\widehat{\mathbf{aBIP}}$ in terms of positive kernels and in terms of the reproducing kernel Hilbert spaces $\mathcal{H}(S)$.

THEOREM 3.1. *Let (1.5) be in force, let P be a positive semidefinite solution of the Lyapunov–Stein equation (1.3), let S be a $p \times q$ mvf which is analytic in $\mathbb{D}$ and let $B(z)$ and $\Lambda_\omega(z)$ be defined by (1.13) and (2.5), respectively. Then the following statements are equivalent:*

(1) *S belongs to the set $\widehat{\mathcal{S}}(M, N, P, C)$.*
(2) *The following kernel is positive on $\mathbb{D} \times \mathbb{D}$:*

$$\mathbf{S}_\omega(z) := \begin{pmatrix} P & B(\omega)^* \\ B(z) & \Lambda_\omega(z) \end{pmatrix} \succeq 0. \tag{3.1}$$

(3) *The matrix $\mathbf{S}_z(z)$ is positive semidefinite for every point $z \in \mathbb{D}$ at which $G(z)$ is invertible.*
(4) *For every choice of $x \in \mathbb{C}^n$, the function $B(z)x$ belongs to the space $\mathcal{H}(S)$ and*

$$\|Bx\|^2_{\mathcal{H}(S)} \leq x^* P x. \tag{3.2}$$

PROOF. $(\mathbf{1}) \Leftrightarrow (\mathbf{2})$. Let $S$ belong to $\widehat{\mathcal{S}}(M, N, P, C)$, let $f_z$ be defined by (2.8) and let

$$h(\zeta) = \begin{pmatrix} C_1 \\ C_2 \end{pmatrix} G(\zeta)^{-1}.$$

Fixing $x \in \mathbb{C}^n$, $y \in \mathbb{C}^p$ and a point $z \in \mathbb{D}$ at which $G(z)$ is invertible, we get

$$\begin{aligned}[] [hy, f_z x]_S &= \left\langle \begin{pmatrix} I_p & -S \\ -S^* & I_q \end{pmatrix} \begin{pmatrix} C_1 \\ C_2 \end{pmatrix} G^{-1} y, \begin{pmatrix} I_p \\ S(z)^* \end{pmatrix} \frac{x}{\rho_z} \right\rangle_{L_2^{p+q}(\mathbb{T})} \\ &= \left\langle (I_p, -S) \begin{pmatrix} C_1 \\ C_2 \end{pmatrix} G^{-1} y, \frac{x}{\rho_z} \right\rangle_{\mathbf{H}_2^p} \\ &= x^* (C_1 - S(z) C_2) G(z)^{-1} y, \end{aligned}$$

in view of (1.8). But this in turn clearly implies that

$$[h, f_z]_S = B(z),$$

since $x$ and $y$ are arbitrary. By Lemma 2.8 and (1.9),

$$[f_\omega, f_z]_S = \Lambda_\omega(z) \quad \text{and} \quad [h, h]_S = P_S \leq P,$$

respectively. Setting $F_\omega(\zeta) = (h(\zeta), f_\omega(\zeta))$ in Lemma 2.9 we conclude that the kernel

$$\widetilde{\mathbf{S}}_\omega(z) := \begin{pmatrix} [h, h]_S & [f_\omega, h]_S \\ [h, f_z]_S & [f_\omega, f_z]_S \end{pmatrix}$$

is positive on $\mathbb{D} \times \mathbb{D}$. Upon substituting the three preceding evaluations into this kernel we come to

$$\mathbf{S}_\omega(z) - \widetilde{\mathbf{S}}_\omega(z) = \begin{pmatrix} P - P_S & 0 \\ 0 & 0 \end{pmatrix} \succeq 0,$$

which implies (3.1).

Now suppose conversely, that (3.1) is in force. Then, setting $z = \omega$ we obtain the matrix inequality

$$\mathbf{S}_z(z) := \begin{pmatrix} P & B(z)^* \\ B(z) & \Lambda_z(z) \end{pmatrix} \succeq 0, \qquad (3.3)$$

which holds for all $z \in \mathbb{D}$, at which $G(z)$ is invertible. This is the *fundamental matrix inequality* (FMI) of the $\widehat{\mathbf{aBIP}}$ and it guarantees that $S$ is a solution of the $\widehat{\mathbf{aBIP}}(M, N, P, C)$; for a proof, see e.g., [**15**, Lemma 3.5].

**(2)** $\Leftrightarrow$ **(3)**. The implication **(2)** $\Rightarrow$ **(3)** is selfevident; the converse is covered by the preceding paragraph.

**(2)** $\Leftrightarrow$ **(4)**. First we note that (3.1) is equivalent to the positivity of the kernels

$$\mathbf{S}_\omega^x(z) := \begin{pmatrix} x^* P x & x^* B(\omega)^* \\ B(z) x & \Lambda_\omega(z) \end{pmatrix} \succeq 0 \qquad (\forall\, x \in \mathbb{C}^n) \qquad (3.4)$$

and thus, it suffices to verify the equivalence of conditions (3.2) and (3.4). If $Px = 0$, then (3.2) is equivalent to $B(z)x \equiv 0$ as well as (3.4). If $Px \neq 0$, then by Schur complements, (3.4) is equivalent to

$$\Lambda_\omega(z) - \gamma^{-1} B(z) x x^* B(\omega)^* \succeq 0, \quad \text{where} \quad \gamma = x^* P x > 0.$$

The rest follows by Proposition 2.5. $\square$

REMARK 3.2. The assumption that $S$ is analytic on $\mathbb{D}$ that is made in the second line of the statement of the Theorem 3.1 can be relaxed. It is enough to assume that $S$ is defined at every point of $\mathbb{D}$. Under this relaxed assumption, statements (1), (2) and (4) in Theorem 3.1 are still equivalent and each one implies statement (3). We shall not pursue this here; see e.g., the discussion of Hindmarsh's theorem in [**21**, p. 36] or the original Hindmarsh paper [**31**] for more on this circle of ideas.

If $S$ is a solution of the $\widehat{\mathbf{aBIP}}$, then the estimate (3.2) for the $\mathcal{H}(S)$–norm of the associated mvf $B$ may be supplemented by the following statement:

LEMMA 3.3. *Let $S \in \mathcal{S}^{p \times q}$ be a solution of the $\widehat{\mathbf{aBIP}}(M, N, P, C)$ and let $B$ be defined by (1.13). Then, for every choice of $x \in \mathbb{C}^n$,*

$$\|Bx\|_{\mathcal{H}(S)}^2 \leq x^* P_S x \qquad (3.5)$$

*with equality if either $S$ is isometric almost everywhere on $\mathbb{T}$ or $C_2 G^{-1} \in \mathbf{H}_2^{q \times n}$.*

PROOF. Taking advantage of characterization (2.6) we get

$$\|Bx\|_{\mathcal{H}(S)}^2 = \sup_{g \in \mathbf{H}_2^q} \left\{ \|Bx + Sg\|_{\mathbf{H}_2^p}^2 - \|g\|_{\mathbf{H}_2^q}^2 \right\}. \qquad (3.6)$$

By (1.10) and (1.13),
$$\begin{aligned}x^*P_Sx &= \|Bx\|_{\mathbf{H}_2^p}^2 + \langle (I_q - S^*S)\, C_2 G^{-1}x,\, C_2 G^{-1}x\rangle_{L_2^q} \\ &= \|Bx\|_{\mathbf{H}_2^p}^2 + \left\|(I_q - S^*S)^{\frac{1}{2}} C_2 G^{-1}x\right\|_{L_2^q}^2,\end{aligned}$$
whereas the bottom condition in (1.8) together with the assumption that $g$ belongs to $\mathbf{H}_2^q$, imply that
$$\begin{aligned}\langle Bx, Sg\rangle_{\mathbf{H}_2^p} &= \langle S^*Bx, g\rangle_{L_2^q} \\ &= \langle S^*(I_p, -S)CG^{-1}x, g\rangle_{L_2^q} \\ &= \langle (I_q - S^*S)\, C_2 G^{-1}x, g\rangle_{L_2^q} - \langle (-S^*, I_q)CG^{-1}x, g\rangle_{L_2^q} \\ &= \langle (I_q - S^*S)\, C_2 G^{-1}x, g\rangle_{L_2^q}.\end{aligned}$$
Therefore,
$$\begin{aligned}\|Bx + Sg\|_{\mathbf{H}_2^p}^2 - \|g\|_{\mathbf{H}_2^q}^2 &= \|Bx\|_{\mathbf{H}_2^p}^2 + 2\Re\langle Bx, Sg\rangle_{\mathbf{H}_2^p} + \|Sg\|_{\mathbf{H}_2^p}^2 - \|g\|_{\mathbf{H}_2^q}^2 \\ &= x^*P_Sx + 2\Re\langle (I_q - S^*S)\,C_2G^{-1}x, g\rangle_{L_2^q} \\ &\quad - \langle (I_q - S^*S)\, C_2 G^{-1}x, C_2 G^{-1}x\rangle_{L_2^q} \\ &\quad - \langle (I_q - S^*S)\,g, g\rangle_{L_2^q} \\ &= x^*P_Sx - \langle (I_q - S^*S)(C_2G^{-1}x - g), C_2G^{-1}x - g\rangle_{L_2^q} \\ &= x^*P_Sx - \left\|(I_q - S^*S)^{\frac{1}{2}}(C_2G^{-1}x - g)\right\|_{L_2^q}^2, \qquad (3.7)\end{aligned}$$
which, in view of (3.6), serves to complete the proof. $\square$

We remark that if $S$ is isometric, then $p \geq q$ and $\mathcal{H}(S) = \mathbf{H}_2^p \ominus S\mathbf{H}_2^q$. For additional information and references, see e.g., [**23**, Section 2].

The next example shows that in general, equality cannot be achieved in (3.5).

EXAMPLE 3.4. Let $M = 0$, $N = 1$, $C_1 = \frac{1}{2}$, $C_2 = 1$ and $P = \frac{3}{4}$. Then $G(z) = -z^{-1}$ and the function $S(z) \equiv \frac{1}{2}$ is a solution of the corresponding $\widehat{\mathbf{aBIP}}$. It follows from (1.10) and (1.13) that for such a choice of the interpolation data,
$$P_S = \frac{3}{4} \quad \text{and} \quad B(z) \equiv 0.$$

The example indicates that we cannot expect equality in (3.5) if $C_2 G^{-1}$ has poles inside $\mathbb{D}$. Nevertheless, the conclusions of the preceding lemma can be strengthened:

LEMMA 3.5. *Let $S \in \mathcal{S}^{p \times q}$ be a solution of the $\widehat{\mathbf{aBIP}}(M, N, P, C)$, let $B$ be defined by (1.13) and let*
$$\det G(z) \neq 0 \quad \text{for} \quad z \in \mathbb{D}. \qquad (3.8)$$
*Then, for every choice of $x \in \mathbb{C}^n$,*
$$\|Bx\|_{\mathcal{H}(S)}^2 = x^*P_Sx.$$

PROOF. Since $S$ is a solution of the $\widehat{\mathbf{aBIP}}(M, N, P, C)$, (3.7) holds. Therefore, by (3.6), the assertion of the lemma is equivalent to

$$\inf_{g \in \mathbf{H}_2^q} \left\| (I_q - S^*S)^{\frac{1}{2}} \left( C_2 G^{-1} x - g \right) \right\|_{L_2^q}^2 = 0.$$

In view of condition (3.8) and Remark 2.12, we may assume without loss of generality that $M = I_n$,

$$N = \begin{pmatrix} A_1 & 0 \\ 0 & A_2 \end{pmatrix}, \quad \text{where} \quad \operatorname{spec} A_1 \subset \mathbb{D}, \quad \operatorname{spec} A_2 \subset \mathbb{T} \qquad (3.9)$$

and the matrices $A_1$ and $A_2$ are in Jordan form. Therefore,

$$G(z) = \begin{pmatrix} G_1(z) & 0 \\ 0 & G_2(z) \end{pmatrix} = \begin{pmatrix} I_{n_1} - zA_1 & 0 \\ 0 & I_{n_2} - zA_2 \end{pmatrix} \quad (n_1 + n_2 = n).$$

Let $C_2 = (C_{21}, C_{22})$ and $x = \begin{pmatrix} x_1 \\ x_2 \end{pmatrix}$ be block decompositions conformal with (3.9) so that

$$C_2 G(z)^{-1} x = C_{21} G_1(z)^{-1} x_1 + C_{22} G_2(z)^{-1} x_2.$$

Then, since $C_{21} G_1^{-1} x_1 \in \mathbf{H}_2^q$,

$$\inf_{g \in \mathbf{H}_2^q} \left\| (I_q - S^*S)^{\frac{1}{2}} \left( C_2 G^{-1} x - g \right) \right\|_{L_2^q}^2 = \inf_{g \in \mathbf{H}_2^q} \left\| (I_q - S^*S)^{\frac{1}{2}} \left( C_{22} G_2^{-1} x_2 - g \right) \right\|_{L_2^q}^2.$$

By another application of (3.7), it is readily seen that the last infimum is equal to

$$(0, x_2^*) P_S \begin{pmatrix} 0 \\ x_2 \end{pmatrix} - \left\| B \begin{pmatrix} 0 \\ x_2 \end{pmatrix} \right\|_{\mathcal{H}(S)}^2.$$

Thus, in order to complete the proof, it suffices to establish the assertion of the lemma for $G(z)$ which is invertible off $\mathbb{T}$. This will be done in Section 14, after the proof of Theorem 14.2. □

The mvf

$$\begin{aligned} W(z) &= -H(z)^{-1} M^* P + H(z)^{-1} C_1^* B(z) \\ &= -H(z)^{-1} M^* P + H(z)^{-1} C_1^* \{C_1 - S(z) C_2\} G(z)^{-1} \end{aligned} \qquad (3.10)$$

plays an important role in the subsequent analysis. It will appear as an entry in the *transformed fundamental matrix inequality* (3.16), which will be obtained from the FMI (3.3) in Theorem 3.8 by invoking an appropriately chosen matrix transformation.

LEMMA 3.6. *Let $P$ be a Hermitian (not necessarily positive semidefinite) solution of the Lyapunov–Stein equation (1.3), let $S$ be a $\mathbb{C}^{p \times q}$-valued function and let $\widetilde{B}$ and $W$ be the mvf's defined in (1.14) and (3.10), respectively. Then*

$$\begin{aligned} C_1^* \left( I_p - S(z) S(\omega)^* \right) C_1 &= H(z) W(z) G(z) + G(\omega)^* W(\omega)^* H(\omega)^* \\ &\quad + \rho_\omega(z) M^* P M + H(z) P H(\omega)^* \\ &\quad - H(z) \widetilde{B}(z) \widetilde{B}(\omega)^* H(\omega)^*. \end{aligned} \qquad (3.11)$$

PROOF. By (3.10) and (1.13),
$$H(z)W(z)G(z) = -M^*PG(z) + C_1^*\left(C_1 - S(z)C_2\right),$$
whereas
$$\rho_\omega(z)M^*PM + H(z)PH(\omega)^* - M^*PG(z) - G(\omega)^*PM = N^*PN - M^*PM$$
$$= C_2^*C_2 - C_1^*C_1,$$
by the definitions (1.4) of $G$, (1.12) of $H$ and the Lyapunov–Stein identity (1.3). Therefore, the expression on the right hand side of (3.11) is equal to
$$C_2^*C_2 - C_1^*C_1 + C_1^*\left(C_1 - S(z)C_2\right) + \left(C_1^* - C_2^*S(\omega)^*\right)C_1$$
$$-(C_2^* - C_1^*S(z))(C_2 - S(\omega)^*C_1) = C_1^*\left(I_p - S(z)S(\omega)^*\right)C_1,$$
which completes the proof of lemma. $\square$

LEMMA 3.7. *Let $P$ be a positive semidefinite solution of the Lyapunov–Stein equation (1.3), let $S$ belong to $\mathcal{S}^{p\times q}$ and let $B$ and $\widetilde{B}$ be the mvf's given by (1.13) and (1.14) respectively. Then the real part of the mvf*
$$\mathbf{W}(z) = zW(z) + \tfrac{1}{2}P = -\tfrac{1}{2}H(z)^{-1}\left(N^* + zM^*\right)P + zH(z)^{-1}C_1^*B(z) \quad (3.12)$$
*is positive semidefinite almost everywhere on $\mathbb{T}$ and*
$$C_1^*\left(I_p - S(z)S(\omega)^*\right)C_1 = H(z)\left(\mathbf{W}(z) + \mathbf{W}(\omega)^* - \widetilde{B}(z)\widetilde{B}(\omega)^*\right)H(\omega)^*$$
$$+\rho_\omega(z)(C_1^*B(z)M + M^*B(\omega)^*C_1 - M^*PM) \quad (3.13)$$
*for every choice of points $z$ and $\omega$ in $\mathbb{D}$.*

PROOF. In view of (3.12) and the formula
$$G(z) = zH(\omega)^* + \rho_\omega(z)M,$$
the expression on the right hand side of (3.11) can be written as
$$H(z)\left(\mathbf{W}(z) + \mathbf{W}(\omega)^* - \widetilde{B}(z)\widetilde{B}(\omega)^*\right)H(\omega)^* + \rho_\omega(z)M^*PM$$
$$+\rho_\omega(z)H(z)W(z)M + \rho_\omega(z)M^*W(\omega)^*H(\omega)^*.$$
Therefore, since
$$H(z)W(z)M = -M^*PM + C_1^*B(z)M,$$
it coincides with the expression on the right hand side of (3.13).

Next, since $\mathbf{W}$, $B$ and $\widetilde{B}$ have boundary values a.e. on $\mathbb{T}$, we may set $z = \omega = e^{it}$ in (3.13). Then, with the help of formulas (1.13)–(1.15), we get
$$\mathbf{W}(e^{it}) + \mathbf{W}(e^{it})^* = G(e^{it})^{-*}C^*\begin{pmatrix} I_p & -S(e^{it}) \\ -S(e^{it})^* & I_q \end{pmatrix}CG(e^{it})^{-1}, \quad (3.14)$$
which shows that $\mathbf{W}$ has positive semidefinite real part a.e. on $\mathbb{T}$, since $S \in \mathcal{S}^{p\times q}$. $\square$

Note that in the formulations of the two preceding lemmas, $S$ is not assumed to be a solution of the $\widehat{\mathbf{aBIP}}$. If $S$ is a solution of the $\widehat{\mathbf{aBIP}}$, then the associated function $\mathbf{W}$ belongs to the Carathéodory class $\mathcal{C}^{n\times n}$ of $n\times n$ mvf's which are analytic and have positive semidefinite real part in $\mathbb{D}$.

THEOREM 3.8. *Let $P$ be a positive semidefinite solution of the Lyapunov-Stein equation (1.3), let $S$ belong to $\widehat{\mathcal{S}}(M, N, P, C)$ and let $B$ and $\widetilde{B}$ be the mvf's given by (1.13), and (1.14), respectively. Then the the mvf $W$ defined by (3.10) is analytic in $\mathbb{D}$ and the transformed kernel*

$$\begin{pmatrix} P & W(\omega)^* \\ W(z) & \dfrac{P + zW(z) + \bar{\omega}W(\omega)^* - \widetilde{B}(z)\widetilde{B}(\omega)^*}{\rho_\omega(z)} \end{pmatrix} \succeq 0 \qquad (3.15)$$

*is positive on $\mathbb{D} \times \mathbb{D}$. In particular, the transformed fundamental matrix inequality*

$$\begin{pmatrix} P & W(z)^* \\ W(z) & \dfrac{P + zW(z) + \bar{z}W(z)^* - \widetilde{B}(z)\widetilde{B}(z)^*}{\rho_z(z)} \end{pmatrix} \geq 0 \qquad (3.16)$$

*holds for every $z \in \mathbb{D}$ and the mvf $\mathbf{W}$ defined by (3.12), belongs to the Carathéodory class $\mathcal{C}^{n \times n}$.*

PROOF. The inequality (3.16) is obtained from the FMI (3.3) by multiplying by

$$E(z) = \begin{pmatrix} G(z)^{-*} & 0 \\ -H(z)^{-1}M^*G(z)^{-*} & H(z)^{-1}C_1^* \end{pmatrix}$$

on the left and by $E(z)^*$ on the right. Details are furnished in the proof of Lemma 3.4 of [15]. The kernel inequality (3.15) may be extracted from the kernel inequality (3.1) in much the same way by multiplying by $E(z)$ on the left and by $E(\omega)^*$ on the right. It follows from (3.16) that

$$\frac{P + zW(z) + \bar{z}W(z)^* - \widetilde{B}(z)\widetilde{B}(z)^*}{\rho_z(z)} = \frac{\mathbf{W}(z) + \mathbf{W}(z)^* - \widetilde{B}(z)\widetilde{B}(z)^*}{\rho_z(z)} \geq 0 \quad (z \in \mathbb{D})$$

and therefore, that $\mathbf{W}(z)$ has a positive semidefinite real part in $\mathbb{D}$. Since, by definition (3.12), $\mathbf{W}(z)$ is meromorphic in $\mathbb{D}$, it is in fact analytic in $\mathbb{D}$ and belongs to $\mathcal{C}^{n \times n}$. □

The *transformed fundamental matrix inequality* was introduced and applied to continuous interpolation problems by V. Katsnelson in [33] and [34]. See also [35], [36] and [38] for various applications of this idea.

The following result will be useful.

LEMMA 3.9. *Let $S$ belong to $\widehat{\mathcal{S}}(M, N, P, C)$ and let the associated mvf $W$ defined via (3.10) belong to $\mathbf{H}_\infty^{n \times n}$. Then $S$ belongs to $\mathcal{S}(M, N, P, C)$.*

PROOF. The assumption on $W$ guarantees that the mvf $\mathbf{W}$ also belongs to $\mathbf{H}_\infty^{n \times n}$ and hence that

$$\frac{1}{2\pi} \int_0^{2\pi} \left( \mathbf{W}(e^{it}) + \mathbf{W}(e^{it})^* \right) dt = \mathbf{W}(0) + \mathbf{W}(0)^* = P.$$

Therefore, by (3.14) and (1.10),

$$P_S := \frac{1}{2\pi} \int_0^{2\pi} G(e^{it})^{-*} C^* \begin{pmatrix} I_p & -S(e^{it}) \\ -S(e^{it})^* & I_q \end{pmatrix} CG(e^{it})^{-1} dt = P.$$

Thus, $S \in \mathcal{S}(M, N, P, C)$. □

## 4. On $\mathcal{H}(\Theta)$ spaces

A $(p+q) \times (p+q)$ mvf $\Theta$ which is meromorphic in $\mathbb{D}$ is said to be *J–inner* in $\mathbb{D}$ if

(1) it is *J*–contractive:
$$\Theta(z) J \Theta(z)^* \leq J$$
at every point $z \in \mathbb{D}$ in the domain of analyticity of $\Theta$ and

(2) its nontangential boundary values, which exist a.e. on $\mathbb{T}$, are *J*–unitary:
$$\Theta(e^{it}) J \Theta(e^{it})^* = J.$$

In this case the kernel
$$K_\Theta(z,\omega) = \frac{J - \Theta(z) J \Theta(\omega)^*}{\rho_\omega(z)} \tag{4.1}$$

is positive on $\mathbb{D} \times \mathbb{D}$ and serves to define a reproducing kernel Hilbert space of $(p+q) \times 1$ meromorphic mvf's in $\mathbb{D}$ which we shall refer to as $\mathcal{H}(\Theta)$. An abstract characterization of reproducing kernel Hilbert spaces of $(p+q) \times 1$ vector valued functions with reproducing kernels of the form (4.1) with a $\Theta$ which is *J*–inner with respect to the upper half plane is due to L. de Branges [16]. His formulation contained an extra technical condition which was later shown to be superfluous by J. Rovnyak [46]. The corresponding characterization for spaces with kernels based on $\Theta$ which are *J*–inner with respect to $\mathbb{D}$ was worked out by J. Ball [11]. A unified approach to both settings and additional generalizations, discussions and references may be found in [4].

All these characterizations are expressed in terms of the generalized backward shift operator
$$(R_\alpha f)(z) = \frac{f(z) - f(\alpha)}{z - \alpha} \tag{4.2}$$
which is defined for every point $\alpha$ in the domain of analyticity of $f$. In order to simplify the discussion, we shall focus on finite dimensional spaces, since these will suffice for our purposes.

THEOREM 4.1. *Let $\mathcal{M}$ be a finite dimensional Hilbert space of $(p+q) \times 1$ vector valued meromorphic functions in $\mathbb{D}$, let $F$ be a $(p+q) \times n$ mvf whose columns form a basis for $\mathcal{M}$ and let $P$ denote the Gram matrix of this basis, i.e.,*
$$\langle Fx, Fy \rangle_\mathcal{M} = y^* P x \tag{4.3}$$
*for every choice of $x$ and $y$ in $\mathbb{C}^n$. Then $\mathcal{M}$ is an $\mathcal{H}(\Theta)$ space for some J–inner mvf $\Theta$ if and only if there exists at least one point $\alpha \in \mathbb{D}$ in the domain of analyticity of $F$ such that*

(1) *$\mathcal{M}$ is $R_\alpha$–invariant and*
(2) *the identity*
$$\langle f, g \rangle_\mathcal{M} + \alpha \langle R_\alpha f, g \rangle_\mathcal{M} + \bar{\alpha} \langle f, R_\alpha g \rangle_\mathcal{M}$$
$$- (1 - |\alpha|^2) \langle R_\alpha f, R_\alpha g \rangle_\mathcal{M} = g(\alpha)^* J f(\alpha) \tag{4.4}$$
*holds for every choice of $f$ and $g$ in $\mathcal{M}$.*

## 4. ON $\mathcal{H}(\Theta)$ SPACES

*Moreover, in this case $F$ is rational and $\Theta$ is uniquely specified up to a right constant $J$–unitary factor by the formula*

$$\Theta(z) = I_{p+q} - \rho_\mu(z) F(z) P^{-1} F(\mu)^* J, \tag{4.5}$$

*where $\mu$ is any point on $\mathbb{T}$ in the domain of analyticity of $F$.*

REMARK 4.2. The $R_\alpha$-invariance imposed in the first condition of Theorem 4.1 forces $F(z)$ to be of the form

$$F(z) = C(M - zN)^{-1} \tag{4.6}$$

for some choice of $C \in \mathbb{C}^{(p+q) \times n}$, $M \in \mathbb{C}^{n \times n}$ and $N \in \mathbb{C}^{n \times n}$ with $M - \alpha N$ invertible (see e.g., [25, Section 3]). Then in fact it turns out that the structural identity (4.4) holds if and only if the Gram matrix $P$ is a solution of the Lyapunov–Stein equation (1.3). The equivalence of the de Branges identity with the Lyapunov–Stein equation for finite dimensional $R_\alpha$-invariant spaces seems to have been first established in [24]; simpler and more transparent proofs may be found in [25] and [27].

Remark 4.2 leads immediately to the following reformulation of Theorem 4.1:

THEOREM 4.3. *Assume that $G(z) = M - zN$ is invertible for at least one point $z \in \mathbb{C}$ and that the columns of $CG(z)^{-1}$ are linearly independent (in the sense that $CG(z)^{-1} x = 0$ for every point $z$ in a nonempty open set in which $G(z)$ is invertible implies that $x$ is the zero vector in $\mathbb{C}^n$). Furthermore, let $P > 0$ be any $n \times n$ positive definite matrix. Then the $n$-dimensional vector space*

$$\mathcal{M} = \{ CG(z)^{-1} x : \quad x \in \mathbb{C}^n \} \tag{4.7}$$

*endowed with the inner product*

$$\langle CG(z)^{-1} x, CG(z)^{-1} y \rangle_\mathcal{M} = y^* P x \tag{4.8}$$

*is an $\mathcal{H}(\Theta)$ space if and only if $P$ is a solution of the Lyapunov–Stein equation (1.3).*

REMARK 4.4. Formulas (4.5) and (4.6) show that if $\mathcal{H}(\Theta)$ is finite dimensional, then $\Theta$ is rational. The converse is also true: $\Theta$ is rational if and only if $\mathcal{H}(\Theta)$ is finite dimensional. Moreover, the McMillan degree of $\Theta$ is equal to the dimension of $\mathcal{H}(\Theta)$. It is perhaps also worth emphasizing that formula (4.5) is a realization formula for $\Theta(z)$. It does not appear in one of the standard forms $D + zC(I_n - zA)^{-1} B$ or $D + C(zI_n - A)^{-1} B$ because it is not "centered" at either zero or infinity. For more information on these and other more general realizations, see [6].

The preceding discussion leads easily to the following useful conclusion:

THEOREM 4.5. *Let $\Theta$ be a rational $J$–inner mvf of McMillan degree $n$. Then there exist matrices $C \in \mathbb{C}^{(p+q) \times n}$, $P \in \mathbb{C}^{n \times n}$, a matrix polynomial $G(z) = M - zN$ with constant coefficients $M$, $N$ in $\mathbb{C}^{n \times n}$ and a point $\mu \in \mathbb{T}$ such that*

(1) *$P$ is positive definite and satisfies the Lyapunov–Stein equation (1.3).*
(2) *$G(\mu)$ is invertible and the columns of $CG(z)^{-1}$ are linearly independent.*
(3) *$\Theta$ is uniquely specified by the formula*

$$\Theta(z) = I_{p+q} - \rho_\mu(z) CG(z)^{-1} P^{-1} G(\mu)^{-*} C^* J, \tag{4.9}$$

*up to a constant $J$–unitary factor on the right (in fact any point $\mu \in \mathbb{T}$ at which $G(\mu)$ is invertible, will do).*

PROOF. By Remark 4.4, $\mathcal{H}(\Theta)$ is an $n$ dimensional space. Let $F$ be a $(p+q) \times n$ mvf whose columns form a basis for $\mathcal{H}(\Theta)$. Then, in view of Theorem 4.1 and Remark 4.2, $F(z) = CG(z)^{-1}$, for some choice of matrices $C$, $M$ and $N$ of the indicated sizes, and $\det G(z) \not\equiv 0$. Moreover, the Gram matrix $P$ that is defined by formula (4.3) for the space $\mathcal{M} = \mathcal{H}(\Theta)$ is a positive semidefinite solution of (1.3), thanks to the de Branges identity (4.4). Formula (4.9) for $\Theta(z)$ emerges by matching the two formulas (4.1) and

$$K_\Theta(z,\omega) = F(z)P^{-1}F(\omega)^*$$

for the reproducing kernel of $\mathcal{H}(\Theta)$ and then choosing $\omega = \mu$, where $\mu$ is a point on $\mathbb{T}$ at which $F$ is analytic, and then further normalizing $\Theta$ by the condition

$$\Theta(\mu) = I_{p+q}. \tag{4.10}$$

□

**Setting and objectives.**

For the rest of this section we shall assume that $C$, $M$, $N$ and $P$ is a given set of matrices such that (1.3)–(1.5) hold and that

the columns of $CG(z)^{-1}$ are linearly independent $\hspace{1em}$ (4.11)

(in the sense explained in the statement of Theorem 4.3) and $P \geq 0$.

Let $r = \operatorname{rank} P$. If $r = n$, i.e., if $P > 0$, then it follows from Theorem 4.3 that the space $\mathcal{M}$ defined by (4.7) and endowed with the inner product (4.3) is an $\mathcal{H}(\Theta)$ space. However, if $r < n$, then (4.8) does not define an inner product. To overcome this difficulty when $r \geq 1$, we shall restrict attention to the subspace

$$\mathcal{M}_Q = \{FQu : \; u \in \mathbb{C}^k\}, \quad \text{where} \quad Q \in \mathbb{C}^{n \times k} \quad \text{and} \quad Q^*PQ > 0. \tag{4.12}$$

Then $k \leq r$ and $\mathcal{M}_Q$ endowed with the inner product (4.3) is a $k$ dimensional reproducing kernel Hilbert space with reproducing kernel

$$K_\omega(z) = F(z)Q(Q^*PQ)^{-1}Q^*F(\omega)^*.$$

However, in order to be an $\mathcal{H}(\Theta)$ space, it is also necessary (and sufficient) that $\mathcal{M}_Q$ is $R_\alpha$–invariant for at least one point $\alpha \in \mathbb{C}$ at which $G(\alpha)$ is invertible.

THEOREM 4.6. *Let $P$ be a positive semidefinite solution of the Lyapunov–Stein equation (1.7), let $\alpha$ be a point in $\mathbb{C}$ at which $G(\alpha)$ is invertible and let (4.11) and (4.12) be in force. Then the following statements are equivalent:*

(1) $\mathcal{M}_Q$ *is $R_\alpha$–invariant.*
(2) $\mathcal{M}_Q$ *is $R_\beta$–invariant for every point $\beta \in \mathbb{C}$ at which $G(\beta)$ is invertible.*
(3) *There exists a $k \times k$ matrix $N_\diamond$ such that*

$$NG(\alpha)^{-1}Q = QN_\diamond.$$

(4) *The vector space $\mathcal{M}_Q$ endowed with the inner product (4.3) is an $\mathcal{H}(\Theta)$ space.*

*Moreover, if any one (and hence everyone) of these conditions is met, then the rational function $\Theta$ of McMillan degree $k$ is uniquely specified by the formula*

$$\Theta(z) = I_{p+q} - \rho_\mu(z) C_\diamond G_\diamond^{-1}(z) P_\diamond^{-1} G_\diamond(\mu)^{-*} C_\diamond^* J \tag{4.13}$$

## 4. ON $\mathcal{H}(\Theta)$ SPACES

up to a J–unitary constant factor on the right, where

$$M_\diamond = I_k + \alpha N_\diamond, \quad G_\diamond(z) = M_\diamond - zN_\diamond, \quad P_\diamond = Q^*PQ, \quad C_\diamond = CG(\alpha)^{-1}Q \tag{4.14}$$

and $\mu$ is any point on $\mathbb{T}$ at which $G(\mu)$ is invertible.

PROOF. The proof is broken into steps.

**Step 1.** *Statements (1) and (3) are equivalent.*

**Proof of Step 1:** Suppose first that (1) holds. Then there exists a $k \times k$ matrix $N_\diamond$ such that

$$(R_\alpha CG^{-1}Q)(z) = CG(z)^{-1}QN_\diamond.$$

But now as

$$(R_\alpha G^{-1})(z) = G(z)^{-1}NG(\alpha)^{-1},$$

it follows that

$$CG(z)^{-1}NG(\alpha)^{-1}Q = CG(z)^{-1}QN_\diamond$$

and hence that (3) holds, since the columns of $CG(z)^{-1}$ are linearly independent. This completes the proof that (1) $\Rightarrow$ (3). The opposite implication drops out easily by running the argument backwards.

**Step 2.** *Let $M_\diamond$ and $G_\diamond(z)$ be defined as in (4.14) and assume that (3) holds. Then $G_\diamond(z)$ is invertible whenever $G(z)$ is invertible and the following formulas are valid:*

$$MG(\alpha)^{-1}Q = QM_\diamond \tag{4.15}$$

*and*

$$G(z)G(\alpha)^{-1}Q = QG_\diamond(z). \tag{4.16}$$

*Moreover,*

$$(R_\beta CG^{-1}Q)(z) = CG(z)^{-1}QN_\diamond G_\diamond(\beta)^{-1} \tag{4.17}$$

*for every point $\beta \in \mathbb{C}$ at which $G(\beta)$ is invertible.*

**Proof of Step 2:** Clearly

$$MG(\alpha)^{-1}Q = (M - \alpha N + \alpha N)G(\alpha)^{-1}Q = Q + \alpha NG(\alpha)^{-1}Q = Q(I_k + \alpha N_\diamond),$$

by (3). This proves (4.15) and leads easily to (4.16). Since $Q$ is a full rank matrix, it follows from (4.16) that $G_\diamond(z)$ is invertible whenever $G(z)$ is invertible and moreover, that

$$G_\diamond(\alpha) = I_k.$$

To obtain (4.17), observe that in view of (4.16),

$$G(z)^{-1}Q = G(\alpha)^{-1}QG_\diamond(z)^{-1}$$

and therefore,

$$\begin{aligned}(R_\beta CG^{-1}Q)(z) &= CG(\alpha)^{-1}Q(R_\beta G_\diamond^{-1})(z) \\ &= CG(\alpha)^{-1}QG_\diamond^{-1}(z)N_\diamond G_\diamond(\beta)^{-1} \\ &= CG(z)^{-1}QN_\diamond G_\diamond(\beta)^{-1},\end{aligned}$$

as claimed.

**Step 3.** *Statements (2) and (3) are equivalent.*

**Proof of Step 3:** By (4.17), (3) $\Rightarrow$ (2). Therefore, since the implication (2) $\Rightarrow$ (1) is selfevident, and (1) $\Rightarrow$ (3) by Step 1, it follows that (2) $\Leftrightarrow$ (3).

**Step 4.** *Statements* (1) *and* (4) *are equivalent and formula* (4.13) *is valid.*

**Proof of Step 4:** If (4) is in force, then, by Theorem 4.1, (1) holds. Conversely, if (1) is in force, then (2) and (3) are also in force and hence, in view of formula (4.16),
$$CG(z)^{-1}Q = CG(\alpha)^{-1}QG_\diamond(z)^{-1} = C_\diamond G_\diamond(z)^{-1}, \tag{4.18}$$
which allows us to characterize $\mathcal{M}_Q$ as
$$\mathcal{M}_Q = \{C_\diamond G_\diamond(z)^{-1}u : \quad u \in \mathbb{C}^k\},$$
endowed with the inner product based on
$$\langle f, g \rangle_{\mathcal{M}_Q} = \langle C_\diamond G_\diamond(z)^{-1}u, C_\diamond G_\diamond(z)^{-1}v \rangle_{\mathcal{M}_Q} = v^* P_\diamond u.$$
It remains to show that $P_\diamond$ satisfies the requisite Lyapunov–Stein identity, By (1.3),
$$Q^* G(\alpha)^{-*} M^* PMG(\alpha)^{-1}Q - Q^* G(\alpha)^{-*} N^* PNG(\alpha)^{-1}Q$$
$$= Q^* G(\alpha)^{-*} C^* JCG(\alpha)^{-1}Q,$$
which, on account of (4.14), can be rewritten as
$$M_\diamond^* P_\diamond M_\diamond - N_\diamond^* P_\diamond N_\diamond = C_\diamond^* JC_\diamond.$$
Since $P_\diamond > 0$, Theorem 4.5 is applicable and thus, $\mathcal{M}_Q = \mathcal{H}(\Theta)$ and $\deg \Theta = k$. □

In this paper we are primarily interested in $R_\alpha$-invariant subspaces $\mathcal{M}_Q$ of dimension $r = \operatorname{rank} P$, since (as will be shown in the next section) these give rise to the $J$-inner function $\Theta$ that is used to describe all the solutions of the $\widehat{\mathbf{aBIP}}$. Consequently, from now on we shall choose $Q \in \mathbb{C}^{n \times r}$ such that
$$Q^* PQ > 0 \quad \text{(and hence} \quad \operatorname{rank} Q^* PQ = \operatorname{rank} P = r) \tag{4.19}$$
and define the subspace
$$\mathcal{M}_Q = \{CG(z)^{-1}Qu : \quad u \in \mathbb{C}^r\}. \tag{4.20}$$
The constraint (4.19) admits the following geometric interpretation: (4.19) holds if and only if the space $\mathbb{C}^n$ can be decomposed as
$$\mathbb{C}^n = \operatorname{Ker} P \dotplus \mathcal{Q} \tag{4.21}$$
where the sum is direct and $\mathcal{Q}$ is the subspace of $\mathbb{C}^n$ given by
$$\mathcal{Q} = \operatorname{Ran} Q := \{Qx : \quad x \in \mathbb{C}^r\}.$$
Furthermore, the constraint (4.19) allows us to define a pseudoinverse matrix
$$P^{[-1]} = Q \left(Q^* PQ\right)^{-1} Q^*, \tag{4.22}$$
which satisfies the equalities
$$P^{[-1]} P P^{[-1]} = P^{[-1]} \quad \text{and} \quad PP^{[-1]}P = P$$
(for a proof, see [15, Lemma 7.1]). Note that in general, the matrix $P^{[-1]}$ defined via (4.22) is not the Moore–Penrose pseudoinverse, since the matrices $P^{[-1]}P$ and $PP^{[-1]}$ are not necessarily Hermitian.

The following theorem is an immediate consequence of Theorem 4.6.

THEOREM 4.7. *Let $P \in \mathbb{C}^{n \times n}$ be a positive semidefinite solution of the Stein equation (1.3) with $\operatorname{rank} P = r$ ($1 \leq r \leq n$), let $\alpha$ be a point in $\mathbb{C}$ at which $G(\alpha)$ is invertible, let $Q$ satisfy (4.19) and assume that (4.11) is in force. Furthermore, let $\mathcal{M}_Q$ be the space defined in (4.20) and endowed with the inner product (4.8). Then $\mathcal{M}_Q = \mathcal{H}(\Theta)$ for some $J$-inner mvf $\Theta$ if and only if*

$$N(M - \alpha N)^{-1} Q = Q N_\diamond \qquad (4.23)$$

*for some $N_\diamond \in \mathbb{C}^{r \times r}$. In this instance, $\Theta$ is a rational mvf of McMillan degree $r$; it can be specified by the formula*

$$\Theta(z) - I_{p+q} - \rho_\mu(z) C G^{-1}(z) P^{[-1]} G(\mu)^{-*} C^* J, \qquad (1.24)$$

*where $\mu \in \mathbb{T}$ is an arbitrary prescribed point at which $G$ is invertible, and is unique up to a $J$-unitary constant factor on the right.*

PROOF. In view of Theorem 4.1, it remains only to show that formulas (4.13) and (4.24) are the same when $k = r$. But this follows readily from (4.18) and (4.22). □

The following example shows that a space $\mathcal{M}$ of the form (4.7) with degenerate inner product may not have an $r$ dimensional subspace which can be identified as an $\mathcal{H}(\Theta)$ space.

EXAMPLE 4.8. Let $p = 2$, $q = 1$, and let

$$P = \begin{pmatrix} 0 & 0 \\ 0 & 1 \end{pmatrix}, \quad M = \begin{pmatrix} 1 & 1 \\ 0 & 1 \end{pmatrix}, \quad N = \begin{pmatrix} 0 & 1 \\ 0 & 0 \end{pmatrix}, \quad C_1 = \begin{pmatrix} 1 & 0 \\ 0 & 1 \end{pmatrix}$$

and $C_2 = (1, 0)$.

Then (1.3) holds, $\operatorname{rank} P = 1$, $G(z) = \begin{pmatrix} 1 & 1-z \\ 0 & 1 \end{pmatrix}$ and $NG(z)^{-1} = \begin{pmatrix} 0 & 1 \\ 0 & 0 \end{pmatrix}$.
Let us assume that (4.23) is satisfied, i.e., that there exist matrices

$$Q = \begin{pmatrix} a \\ b \end{pmatrix} \in \mathbb{C}^{2 \times 1} \quad \text{and} \quad N_\diamond \in \mathbb{C}^{1 \times 1}$$

such that

$$Q^* P Q = |b|^2 > 0 \quad \text{and} \quad \begin{pmatrix} b \\ 0 \end{pmatrix} = NG(z)^{-1} Q = Q N_\diamond = \begin{pmatrix} a N_\diamond \\ b N_\diamond \end{pmatrix}. \qquad (4.25)$$

The second relation in (4.25) implies that $b N_\diamond = 0$ and $b = a N_\diamond$, which is not compatible with the first relation in (4.25).

## 5. Parametrizations of all solutions

As we have already mentioned, the $\widehat{\mathbf{aBIP}}$ may be considered using Potapov's method of fundamental matrix inequalities. In particular, by Theorem 3.1, the set $\widehat{\mathcal{S}}(M, N, P, C)$ is the same as the set of all solutions $S$ of the FMI (3.3). This set was parametrized in [15] in terms of a linear fractional transformation under the additional assumption that there exists a matrix $Q \in \mathbb{C}^{n \times r}$ that meets (4.19) such that

$$M\mathcal{Q} \subseteq \mathcal{Q} \quad \text{and} \quad N\mathcal{Q} \subseteq \mathcal{Q}, \qquad (5.1)$$

where the $r$ dimensional subspace $\mathcal{Q}$ is equal to the range of $Q$.

A decomposition of the form (4.21) with the above mentioned invariance properties was used by V. Dubovoj in [**22**] to study the degenerate matrix Schur problem. This corresponds to the case where $M = I_n$ and $N$ is equal to the block shift matrix. In more general settings, however, a decomposition of this form with the requisite invariance properties may not exist.

The invariance properties of the subspace $\mathcal{Q}$ enable us to build a linear fractional transformation describing all the solutions of the degenerate $\widehat{\mathbf{aBIP}}(M, N, P, C)$. The matrix of coefficients of the linear fractional transformation

$$\Theta = \begin{pmatrix} \theta_{11} & \theta_{12} \\ \theta_{21} & \theta_{22} \end{pmatrix} : \begin{pmatrix} \mathbb{C}^p \\ \mathbb{C}^q \end{pmatrix} \to \begin{pmatrix} \mathbb{C}^p \\ \mathbb{C}^q \end{pmatrix} \quad (5.2)$$

is given by (4.24); it is $J$-inner in $\mathbb{D}$, by Theorem 4.7.

THEOREM 5.1. *Let (4.11), (4.19) and (5.1) (or (4.23)) be in force, let $P \in \mathbb{C}^{n \times n}$ be a positive semidefinite solution of the Lyapunov–Stein equation (1.3) of rank $r$ (where $1 \le r \le n$) and let the mvf $\Theta$ given by (4.24) be decomposed into four blocks as in (5.2). Then all the solutions $S$ of the $\widehat{\mathbf{aBIP}}(M, N, P, C)$ are parametrized by the linear fractional transformation*

$$S(z) = \mathbf{T}_\Theta(\mathcal{E}) := (\theta_{11}(z)\mathcal{E}(z) + \theta_{12}(z)) (\theta_{21}(z)\mathcal{E}(z) + \theta_{22}(z))^{-1}, \quad (5.3)$$

*in which the parameter $\mathcal{E} \in \mathcal{S}^{p \times q}$ is of the form*

$$\mathcal{E}(z) = U \begin{pmatrix} \widehat{\mathcal{E}}(z) & 0 \\ 0 & I_\nu \end{pmatrix} V, \quad (5.4)$$

*where $U \in \mathbb{C}^{p \times p}$ and $V \in \mathbb{C}^{q \times q}$ are fixed unitary matrices that depend only on the interpolation data,*

$$\nu = \mathrm{rank}\,(M^*PM + C_2^*C_2) - \mathrm{rank}\,P = \mathrm{rank}\,(N^*PN + C_1^*C_1) - \mathrm{rank}\,P, \quad (5.5)$$

*and $\widehat{\mathcal{E}}(z)$ is an arbitrary mvf in the Schur class $\mathcal{S}^{(p-\nu) \times (q-\nu)}$.*

PROOF. In view of Theorem 3.1, it suffices to show that the transformation (5.3) with $\mathcal{E}$ as in (5.4) parametrizes all the solutions of the FMI (3.1). But this is precisely what is shown in the proof of Theorem 7.4 of [**15**]. □

REMARK 5.2. The case $r = 0$, is not covered by Theorem 5.1. However, this case is simple: if $S$ is a solution of the $\widehat{\mathbf{aBIP}}(M, N, P, C)$ with $r = 0$, then $P = 0$, $P_S = 0$, $S(z)C_2 \equiv C_1$ and the Stein equation (1.3) implies that $C_1^*C_1 = C_2^*C_2$. Therefore, $S$ is of the form $U \begin{pmatrix} s(z) & 0 \\ 0 & I_\nu \end{pmatrix} V$, where $\nu = \mathrm{rank}\,C_1 = \mathrm{rank}\,C_2$ and $U$ and $V$ are unitary; see e.g., [**23**, p.8] for help with this, if need be. For the rest of this paper we shall focus on the case $r \ge 1$.

Note that for the nondegenerate $\widehat{\mathbf{aBIP}}(M, N, P, C)$ ($\det P \ne 0$) the integer $\nu$ defined by (5.5) is equal to zero; therefore, there is no constant block in (5.4) and the parameter $\mathcal{E}$ varies over all of $\mathcal{S}^{p \times q}$. Moreover, the function $\Theta$ is now given by formula (4.9). Thus, for the case when $\det P \ne 0$ the last theorem simplifies as follows.

THEOREM 5.3. *Let $P \in \mathbb{C}^{n \times n}$ be a positive definite solution of the Lyapunov–Stein equation (1.3), let (4.11) be in force and let the mvf $\Theta$ given by (4.9) be decomposed into four blocks as in (5.2). Then all the solutions $S$ of the $\widehat{\mathbf{aBIP}}(M, N, P, C)$*

# 5. PARAMETRIZATIONS OF ALL SOLUTIONS

are parametrized by the linear fractional transformation (5.3), in which the parameter $\mathcal{E}$ varies over all of $\mathcal{S}^{p\times q}$.

The next theorem characterizes the set $\widehat{\mathcal{S}}(M, N, P, C)$ when $P$ is a singular positive semidefinite solution of the Stein equation (1.3) of rank $r \geq 1$.

THEOREM 5.4. *Let $P \in \mathbb{C}^{n\times n}$ be a positive semidefinite solution of the Lyapunov–Stein equation (1.3), let (4.11) be in force and let (4.19) and (5.1) be in force for some matrix $Q \in \mathbb{C}^{n\times r}$, $M_\diamond$ and $N_\diamond$ and let $\Theta$ be defined as in (4.24). Then a mvf $S$ is of the form (5.3) for some $\mathcal{E} \in \mathcal{S}^{p\times q}$ of the form (5.4) if and only if it is a solution of the $\widehat{\mathbf{aBIP}}(M_\diamond, N_\diamond, P_\diamond, C_\diamond)$, where $M_\diamond$, $N_\diamond$, $P_\diamond$ and $C_\diamond$ are the matrices defined in (4.14).*

The proof follows immediately from Theorem 5.3 and the representation (4.13) of $\Theta$.

Theorem 5.4 is not applicable if condition (5.1) is not in force. If $G(z)$ is invertible on $\mathbb{T}$, then some of the off diagonal entries in $M$ and $N$ can be modified as in [**15**, Section 5] to produce a new problem that has the same set of solutions as the given problem and for which (5.1) is in force. However, in the present setting it is more convenient to parametrize the set of all solutions of the $\widehat{\mathbf{aBIP}}$ in terms of a Redheffer linear fractional transformation. Accordingly, we now recall some facts from [**15**] on a Redheffer type representation for the set $\widehat{\mathcal{S}}(M, N, P, C)$ of all the solutions of the $\widehat{\mathbf{aBIP}}$ based on the methods of [**36**]. Let us consider the $n \times n$ matrix valued function

$$\Delta_\omega(z) = G(\omega)^* P G(z) + \rho_\omega(z) C_2^* C_2. \tag{5.6}$$

which was introduced in [**26**] and used extensively in [**26**] and [**27**] (in a more general setting) and which will play an important role in this paper too. We begin with a list of formulas which are taken from [**26**] and [**27**]. They can be verified by straightforward computation, especially if they are tackled in the order in which they are stated.

LEMMA 5.5. *If $P \geq 0$ is a solution of the Lyapunov–Stein equation (1.7), then the following formulas are valid for every choice of $z$ and $\omega$ in $\mathbb{C}$:*

(1) $\Delta_\omega(z) = H(z) P H(\omega)^* + \rho_\omega(z) C_1^* C_1,$ \hfill (5.7)

(2) $\Delta_\omega(z) = \Delta_z(\omega)^*,$

(3) $\rho_\omega(z) G(z)^* P G(\omega) + \rho_\omega(z)^* H(z) P H(\omega)^*$
$= \rho_z(z) G(\omega)^* P G(\omega) + \rho_\omega(\omega) H(z) P H(z)^*,$

(4) $\rho_\omega(z)^* \Delta_\omega(z) + \rho_\omega(z) \Delta_\omega(z)^*$
$= \rho_z(z) G(\omega)^* P G(\omega) + \rho_\omega(\omega) H(z) P H(z)^* + |\rho_\omega(z)|^2 \left(C_1^* C_1 + C_2^* C_2\right).$

It turns out that the kernel of the matrix $\Delta_\omega(z)$ does not depend on the choice of $z$ and $\omega$ in $\mathbb{D}$. To be more precise, let us introduce the subspace

$$\mathcal{K} = \operatorname{Ker} PM \,\cap\, \operatorname{Ker} PN \,\cap\, \operatorname{Ker} C \tag{5.8}$$

and a family of subspaces
$$\mathcal{K}_\zeta = \operatorname{Ker} PG(\zeta) \cap \operatorname{Ker} C \qquad (\zeta \in \mathbb{T}). \tag{5.9}$$

The formulas
$$\mathcal{K} = \operatorname{Ker} PM \cap \operatorname{Ker} C_2 = \operatorname{Ker} PN \cap \operatorname{Ker} C_1,$$
which follow readily from the identity
$$M^*PM + C_2^*C_2 = N^*PN + C_1^*C_1$$
(which is just another way of writing (1.7)), allow us to express the integer $\nu$ in (5.5) in terms of $\mathcal{K}$ as
$$\nu = \dim \operatorname{Ker} P - \dim \mathcal{K}.$$
The inclusion $\mathcal{K} \subseteq \mathcal{K}_\zeta$ is selfevident; equality prevails for every point $\zeta \in \mathbb{T}$ at which $G(\zeta)$ is invertible. This is an easy consequence of the identity
$$G(\zeta)^*PM + \bar\zeta N^*PG(\zeta) = C^*JC,$$
which is valid for $\zeta \in \mathbb{T}$ and solutions $P$ of the Lyapunov–Stein equation (1.7). This inclusion is also established in the next lemma, which extends Lemma 6.2 of [15]. The extension (i.e., (5.11)) is an easy consequence of Lemma 5.5. The third statement is immediate from (5.10) and (5.11).

LEMMA 5.6. *Let $P \geq 0$ be a solution of the Lyapunov–Stein equation (1.7) and let $\omega \in \mathbb{D}$. Then:*

(1) *For every point $z \in \mathbb{D}$ and for every point $z \in \mathbb{T}$ at which $G(z)$ is invertible,*
$$\operatorname{Ker} \Delta_\omega(z) = \operatorname{Ker} \Delta_\omega(z)^* = \mathcal{K}. \tag{5.10}$$

(2) *For every point $\zeta \in \mathbb{T}$,*
$$\operatorname{Ker} \Delta_\omega(\zeta) = \operatorname{Ker} \Delta_\omega(\zeta)^* = \mathcal{K}_\zeta. \tag{5.11}$$

(3) *The subspace $\mathcal{K}_\zeta$ coincides with $\mathcal{K}$ for every point $\zeta \in \mathbb{T}$ at which $G(\zeta)$ is invertible.*

From now on, let $\dim \mathcal{K} = k$ and let $\mathbf{Q} \in \mathbb{C}^{n \times (n-k)}$ be an isometric matrix whose columns span $\mathcal{K}^\perp$, the orthogonal complement of $\mathcal{K}$ in $\mathbb{C}^n$ with respect to the standard inner product.

The next conclusion is a consequence of the preceding two lemmas.

LEMMA 5.7. *If $P \geq 0$ is a solution of the Lyapunov–Stein equation (1.7) and $\omega \in \mathbb{D}$, then $\mathbf{Q}^*\Delta_\omega(z)\mathbf{Q}$ is invertible for every point $z \in \mathbb{D}$ and for every point $z \in \mathbb{T}$ at which $G(z)$ is invertible. Moreover, if*
$$\mathcal{K}_\zeta = \mathcal{K} \quad \text{for every point} \quad \zeta \in \mathbb{T} \quad \text{at which} \quad \det G(\zeta) = 0, \tag{5.12}$$
*then $\mathbf{Q}^*\Delta_\omega(z)\mathbf{Q}$ is invertible in the closed unit disk $\overline{\mathbb{D}}$.*

We now define
$$\Delta_\omega^{[-1]}(z) := \mathbf{Q}\left(\mathbf{Q}^*\Delta_\omega(z)\mathbf{Q}\right)^{-1}\mathbf{Q}^*,$$
for all points $z \in \mathbb{C}$ at which the indicated inverse exists. Since $\det\{\mathbf{Q}^*\Delta_\omega(z)\mathbf{Q}\}$ is a polynomial in $z$ of degree at most $n-k$, which has no zeros in $\mathbb{D}$ by the preceding lemma, the inverse can fail to exist at most at $n-k$ points, all of which fall outside $\mathbb{D}$. Moreover, Lemma 5.6 guarantees that:

LEMMA 5.8. *If $P \geq 0$ is a solution of the Lyapunov–Stein equation (1.7) and $\omega \in \mathbb{D}$, then the function $\Delta_\omega^{[-1]}(z)$ is rational and has at most $n - k$ poles all of which fall outside $\mathbb{D}$. Moreover, if the condition (5.12) is in force, then $\Delta_\omega^{[-1]}(z)$ is analytic in the closed unit disk $\overline{\mathbb{D}}$.*

Let $P \geq 0$ be a solution of the Lyapunov–Stein equation (1.7), let $\omega \in \mathbb{D}$ and let $\mathbf{P}_{\mathcal{K}}$ denote the orthogonal projection of $\mathbb{C}^n$ onto $\mathcal{K}$. Then the following equalities hold at every point $z \in \overline{\mathbb{D}}$ at which $\Delta_\omega^{[-1]}(z)$ is analytic (for a proof, see [**15**, Lemma 6.9]):

(1) $\Delta_\omega^{[-1]}(z) \Delta_\omega(z) \Delta_\omega^{[-1]}(z) = \Delta_\omega^{[-1]}(z)$.

(2) $\Delta_\omega(z) \Delta_\omega^{[-1]}(z) \Delta_\omega(z) = \Delta_\omega(z)$.

(3) $\Delta_\omega(z) \Delta_\omega^{[-1]}(z) = \Delta_\omega^{[-1]}(z) \Delta_\omega(z) = I_n - \mathbf{P}_{\mathcal{K}}$. (5.13)

These equalities serve to show that $\Delta_\omega^{[-1]}(z)$ is the Moore–Penrose pseudoinverse (see e.g. [**42**, Section 12.8]) of $\Delta_\omega(z)$ for every point $z \in \overline{\mathbb{D}}$ at which $\Delta_\omega^{[-1]}(z)$ is analytic.

Next, following the analysis in Section 12 of [**27**] and its refinement in Section 10 of [**15**], let

$$W_1 = \begin{pmatrix} P^{\frac{1}{2}} G(\omega) \\ \rho_\omega(\omega)^{\frac{1}{2}} C_2 \end{pmatrix} \quad \text{and} \quad W_2 = \begin{pmatrix} -P^{\frac{1}{2}} H(\omega)^* \\ \rho_\omega(\omega)^{\frac{1}{2}} C_1 \end{pmatrix} \quad (5.14)$$

for some fixed choice of $\omega \in \mathbb{D}$. Evaluating (5.6) and (5.7) at the point $\omega$ we get

$$\Delta_\omega(\omega) = G(\omega)^* P G(\omega) + \rho_\omega(\omega) C_2^* C_2 = H(\omega) P H(\omega)^* + \rho_\omega(\omega) C_1^* C_1,$$

which can be written as

$$\Delta_\omega(\omega) = W_1^* W_1 = W_2^* W_2 \quad (5.15)$$

and guarantees that the linear map

$$\mathbf{V} : W_1 x \longrightarrow W_2 x$$

is an isometry from $\mathcal{D}_{\mathbf{V}} = \operatorname{Ran} W_1 \subset \mathbb{C}^{n+q}$ onto $\mathcal{R}_{\mathbf{V}} = \operatorname{Ran} W_2 \subset \mathbb{C}^{n+p}$. By (5.15),

$$\dim \mathcal{D}_{\mathbf{V}} = \dim \mathcal{R}_{\mathbf{V}} = \operatorname{rank} \Delta_\omega(\omega) = n - k,$$

and thus, the dimensions of the orthogonal complements

$$\mathcal{D}_{\mathbf{V}}^{\perp} = \mathbb{C}^{n+q} \ominus \mathcal{D}_{\mathbf{V}} \quad \text{and} \quad \mathcal{R}_{\mathbf{V}}^{\perp} = \mathbb{C}^{n+p} \ominus \mathcal{R}_{\mathbf{V}}$$

are equal to

$$q' := \dim \mathcal{D}_{\mathbf{V}}^{\perp} = k + q \quad \text{and} \quad p' := \dim \mathcal{R}_{\mathbf{V}}^{\perp} = k + p,$$

respectively. Let $W_1^{\perp} \in \mathbb{C}^{(n+q) \times q'}$ and $W_2^{\perp} \in \mathbb{C}^{(n+p) \times p'}$ be isometric matrices whose columns span $\mathcal{D}_{\mathbf{V}}^{\perp}$ and $\mathcal{R}_{\mathbf{V}}^{\perp}$, respectively. Since $\operatorname{rank} P = r$, it follows from (5.14) that $W_1^{\perp}$ and $W_2^{\perp}$ can be chosen in the form

$$W_1^{\perp} = \begin{pmatrix} X & Y_1 \\ 0 & Z_1 \end{pmatrix} \quad \text{and} \quad W_2^{\perp} = \begin{pmatrix} X & Y_2 \\ 0 & Z_2 \end{pmatrix}, \quad (5.16)$$

where $X \in \mathbb{C}^{n \times (n-r)}$ is an isometric matrix whose columns form an orthonormal basis for $\operatorname{Ker} P$. The main conclusion is formulated below as Theorem 5.9. The formulas that are exhibited in Theorem 5.9 are derived in [**15**, Theorem 10.1]; an earlier version with some additional restrictions appears as [**27**, Theorem 12.1].

The fact that the set of all solutions of the $\widehat{\mathbf{aBIP}}$ is parametrized by the Redheffer transformation (5.17) rests on the results of [**36**]). The analysis in [**36**] is applicable because the set of all solutions of the $\widehat{\mathbf{aBIP}}(M, N, P, C)$ coincides with the set of solutions to the **AIP** problem that is formulated in [**36**]) when it is specialized to the setting of the $\widehat{\mathbf{aBIP}}(M, N, P, C)$.

THEOREM 5.9. *Let $P \geq 0$ be a solution of the Lyapunov–Stein equation (1.7), let $\omega \in \mathbb{D}$, let $\Delta_\omega(z)$ be the function defined in (5.6) and let $\delta_\omega(z) = z - \omega$. Then all the solutions $S$ of the $\widehat{\mathbf{aBIP}}$ are parametrized by the Redheffer transformation*

$$S(z) = \Psi_{12}(z) + \Psi_{11}(z)\mathcal{E}(z)\left(I_q - \Psi_{21}(z)\mathcal{E}(z)\right)^{-1}\Psi_{22}(z), \qquad (5.17)$$

*where*

$$\Psi_{11}(z) = Z_2 + \rho_\omega(\omega)^{-\frac{1}{2}}\delta_\omega(z)C_1\Delta_\omega^{[-1]}(z)G(\omega)^*P^{\frac{1}{2}}Y_2, \qquad (5.18)$$

$$\Psi_{12}(z) = \rho_\omega(z)C_1\Delta_\omega^{[-1]}(z)C_2^*, \qquad (5.19)$$

$$\Psi_{21}(z) = \frac{\delta_\omega(z)}{\rho_\omega(z)}Y_1^*\left(I - \frac{\delta_\omega(z)}{\rho_\omega(\omega)}P^{\frac{1}{2}}H(\omega)^*\Delta_\omega^{[-1]}(z)G(\omega)^*P^{\frac{1}{2}}\right)Y_2, \qquad (5.20)$$

$$\Psi_{22}(z) = Z_1^* - \rho_\omega(\omega)^{-\frac{1}{2}}\delta_\omega(z)Y_1^*P^{\frac{1}{2}}H(\omega)^*\Delta_\omega^{[-1]}(z)C_2^*, \qquad (5.21)$$

*$\nu$ is the number given by (5.5) and $\mathcal{E}$ is a free independent parameter varying over $\mathcal{S}^{(p-\nu) \times (q-\nu)}$. The mvf*

$$\Psi(z) = \begin{pmatrix} \Psi_{11}(z) & \Psi_{12}(z) \\ \Psi_{21}(z) & \Psi_{22}(z) \end{pmatrix}$$

*is inner in $\mathbb{D}$.*

REMARK 5.10. *If $\omega = 0$, then the formulas (5.18)–(5.21) simplify to*

$$\Psi_{11}(z) = Z_2 + zC_1\Delta_0^{[-1]}(z)M^*P^{\frac{1}{2}}Y_2, \qquad (5.22)$$

$$\Psi_{12}(z) = C_1\Delta_0^{[-1]}(z)C_2^*, \qquad (5.23)$$

$$\Psi_{21}(z) = zY_1^*\left(I + zP^{\frac{1}{2}}N\Delta_0^{[-1]}(z)M^*P^{\frac{1}{2}}\right)Y_2, \qquad (5.24)$$

$$\Psi_{22}(z) = Z_1^* + zY_1^*P^{\frac{1}{2}}N\Delta_0^{[-1]}(z)C_2^*, \qquad (5.25)$$

*where*

$$\Delta_0^{[-1]}(z) = \mathbf{Q}\left(\mathbf{Q}^*\{M^*P(M-zN) + C_2^*C_2\}\mathbf{Q}\right)^{-1}\mathbf{Q}^* \qquad (5.26)$$

*and $\mathbf{Q} \in \mathbb{C}^{n \times (n-k)}$ is an isometric matrix whose columns span the orthogonal complement of the subspace $\mathcal{K}$ of $\mathbb{C}^n$ that is defined by (5.8) with respect to the standard inner product in $\mathbb{C}^n$.*

COROLLARY 5.11. *The $\widehat{\mathbf{aBIP}}(M, N, P, C)$ has a unique solution (which is rational) if and only if $\nu = \min(p, q)$ and it has infinitely many solutions if $\nu < \min(p, q)$.*

This conclusion follows immediately from the parametrization (5.17) of all the solutions of the $\widehat{\mathbf{aBIP}}$: if $\nu = \min(p, q)$, then the second term does not appear on the right hand side of (5.17) and the mvf $\Psi_{12}$ (which is clearly rational) is the unique solution of the $\widehat{\mathbf{aBIP}}$. It turns out that even for the indeterminate case, $\Psi_{12}$ is a very special solution of the $\widehat{\mathbf{aBIP}}$ with important extremal properties.

The next lemma summarizes some useful formulas that will be used in the sequel. A justification of these formulas and another related formula is furnished in the proof of Lemmas 10.2 and 10.3 of [**15**].

LEMMA 5.12. *Let $P \geq 0$ be a solution of the Lyapunov–Stein equation (1.7), let $z, \omega \in \mathbb{D}$ and let $\Psi_{12}$, $\Psi_{11}$ and $\Psi_{22}$ be the functions given by (5.19), (5.18) and (5.21), respectively. Then*

$$(I_p, -\Psi_{12}(z))\, CG^{-1}(z) = C_1 \Delta_\omega^{[-1]}(z) G(\omega)^* P, \tag{5.27}$$

$$(-\Psi_{12}(z)^*, I_q)\, CH(z)^{-*} = C_2 \Delta_\omega^{[-1]}(z)^* H(\omega) P, \tag{5.28}$$

$$C_1^* \Psi_{11}(z) = H(z) P^{\frac{1}{2}} \Upsilon_\omega(z) Y_2, \tag{5.29}$$

$$\Psi_{22}(z) C_2 = -Y_1^* \Upsilon_\omega(z) P^{\frac{1}{2}} G(z), \tag{5.30}$$

*where*

$$\Upsilon_\omega(z) = \frac{\rho_\omega(\omega)^{\frac{1}{2}}}{\rho_\omega(z)} \left\{ I - \frac{\delta_\omega(z)}{\rho_\omega(\omega)} P^{\frac{1}{2}} H(\omega)^* \Delta_\omega^{[-1]}(z) G(\omega)^* P^{\frac{1}{2}} \right\}. \tag{5.31}$$

## 6. The equality case

In this section we return to a general **aBIP**$(M, N, P, C)$ and establish necessary and sufficient conditions for this problem to be solvable.

THEOREM 6.1. *Let $P$ be a positive semidefinite solution of the Lyapunov–Stein equation (1.3) and let $\mathcal{K}$ and $\mathcal{K}_\zeta$ be the subspaces defined in (5.8) and (5.9), respectively. Then the **aBIP**$(M, N, P, C)$ is solvable if and only if $\mathcal{K} = \mathcal{K}_\zeta$ for every point $\zeta \in \mathbb{T}$ at which $G(\zeta)$ is not invertible (i.e., if and only if (5.12) holds[3]).*

PROOF. Since $P$ is a positive semidefinite solution of the Lyapunov–Stein equation (1.3), the $\widehat{\mathbf{aBIP}}(M, N, P, C)$ is solvable and all its solutions are parametrized by formula (5.17). In particular, the function $\Psi_{12}(z)$ defined in (5.19) is a solution of the $\widehat{\mathbf{aBIP}}(M, N, P, C)$ (corresponding to the parameter $\widehat{\mathcal{E}}(z) \equiv 0$). We show that under assumption (5.12), $\Psi_{12}(z)$ is a solution of **aBIP**$(M, N, P, C)$. Let $W_0$ be the mvf constructed via (3.10) from $\Psi_{12}$:

$$W_0(z) = -H(z)^{-1} M^* P + H(z)^{-1} C_1^* \begin{pmatrix} I_p, & -\Psi_{12}(z) \end{pmatrix} CG(z)^{-1}.$$

Then by (5.27),

$$W_0(z) = -H(z)^{-1} M^* P + H(z)^{-1} C_1^* C_1 \Delta_\omega^{[-1]}(z) G(\omega)^* P. \tag{6.1}$$

Making use of (5.7) and (5.13), we get

$$\rho_\omega(z) C_1^* C_1 \Delta_\omega^{[-1]}(z) = I_n - \mathbf{P}_\mathcal{K} - H(z) P H(\omega)^* \Delta_\omega^{[-1]}(z), \tag{6.2}$$

where $\mathbf{P}_\mathcal{K}$ is the orthogonal projection of $\mathbb{C}^n$ onto $\mathcal{K}$. By definition (5.8),

$$P G(\omega) \mathbf{P}_\mathcal{K} = 0$$

and therefore, upon multiplying (6.2) by $G(\omega)^* P$ on the right, we get

$$\rho_\omega(z) C_1^* C_1 \Delta_\omega^{[-1]}(z) G(\omega)^* P = G(\omega)^* P - H(z) P H(\omega)^* \Delta_\omega^{[-1]}(z) G(\omega)^* P.$$

---

[3]We remark that in view of Lemma 5.5, the condition (5.12) holds if and only if $\mathcal{K}_\zeta = \mathcal{K}$ for every point $\zeta \in \mathbb{T}$.

Substituting the latter relation into (6.1) and making use of the equality
$$G(\omega)^* - \rho_\omega(z)M^* = \bar{\omega}H(z),$$
we obtain
$$W_0(z) = \frac{\bar{\omega}}{\rho_\omega(z)}P - \frac{1}{\rho_\omega(z)}PH(\omega)^*\Delta_\omega^{[-1]}(z)G(\omega)^*P. \tag{6.3}$$

Let $\mathcal{K} = \mathcal{K}_\zeta$ for every point $\zeta \in \mathbb{T}$ at which $\det G(\zeta) = 0$. Then by Lemma 5.8, the rational mvf $\Delta_\omega^{[-1]}(z)$ is analytic in the closed unit disk $\overline{\mathbb{D}}$ and it follows readily from (6.3) that the rational mvf $W_0$ is analytic in $\overline{\mathbb{D}}$ as well. Therefore, $\Psi_{12}$ is a solution of $\mathbf{aBIP}(M, N, P, C)$ by Lemma 3.9. This completes the proof of the sufficiency of (5.12).

Suppose next that $S$ is a solution of the $\widehat{\mathbf{aBIP}}(M, N, P, C)$, that $\zeta_0 \in \mathbb{T}$ and that $f \in \mathcal{K}_{\zeta_0}$, i.e.,
$$PG(\zeta_0)f = 0 \quad \text{and} \quad Cf = 0. \tag{6.4}$$

By Theorem 3.8, the mvf $\mathbf{W}$ defined by (3.12) belongs to Carathéodory class $\mathcal{C}^{n \times n}$ and the inequality (3.16) holds for all $z \in \mathbb{D}$. Multiplying (3.16) by the matrix $\begin{pmatrix} G(\zeta_0)f & 0 \\ 0 & I_p \end{pmatrix}$ on the right, by its adjoint on the left and taking advantage of the first relation in (6.4), we conclude that
$$W(z)G(\zeta_0)f \equiv 0.$$

Consequently, the mvf $\mathbf{W}$ which is defined by (3.12), satisfies
$$\mathbf{W}(z)G(\zeta_0)f = \left(zW(z) + \tfrac{1}{2}P\right)G(\zeta_0)f \equiv 0.$$

Therefore,
$$f^*G(\zeta_0)^*\left(\mathbf{W}(\zeta) + \mathbf{W}(\zeta)^*\right)G(\zeta_0)f = 0$$
at almost every point $\zeta \in \mathbb{T}$. In view of (3.14), the last equality can be written as
$$f^*G(\zeta_0)^*G(\zeta)^{-*}C^*\begin{pmatrix} I_p & -S(\zeta) \\ -S(\zeta)^* & I_q \end{pmatrix}CG(\zeta)^{-1}G(\zeta_0)f = 0. \tag{6.5}$$

Since
$$G(\zeta)^{-1}G(\zeta_0) = I_n + (\zeta - \zeta_0)G(\zeta)^{-1}N$$
and $Cf = 0$, it follows from (6.5) that
$$|\zeta - \zeta_0|^2 f^*N^*G(\zeta)^{-*}C^*\begin{pmatrix} I_p & -S(\zeta) \\ -S(\zeta)^* & I_q \end{pmatrix}CG(\zeta)^{-1}Nf = 0.$$

Therefore,
$$f^*N^*G(\zeta)^{-*}C^*\begin{pmatrix} I_p & -S(\zeta) \\ -S(\zeta)^* & I_q \end{pmatrix}CG(\zeta)^{-1}Nf = 0$$
for a.e. $\zeta \in \mathbb{T}$ and hence, by (1.10),
$$\begin{aligned} f^*N^*P_SNf &= \frac{1}{2\pi}\int_0^{2\pi} f^*N^*G(e^{it})^{-*}C^*\begin{pmatrix} I_p & -S(e^{it}) \\ -S(e^{it})^* & I_q \end{pmatrix}CG(e^{it})Nf\,dt \\ &= 0. \end{aligned}$$

Thus, $P_SNf = 0$ for every solution $S$ of the $\widehat{\mathbf{aBIP}}(M, N, P, C)$ and hence, if $P_S = P$, then it is readily seen that $f \in \mathcal{K}$. This proves that $\mathcal{K}_{\zeta_0} \subseteq \mathcal{K}$ and thus, as the opposite inclusion is selfevident and $\zeta_0$ is an arbitrary point on the unit circle, $\mathcal{K}_\zeta = \mathcal{K}$ for every $\zeta \in \mathbb{T}$. $\square$

As a byproduct of this analysis, we obtain the following conclusion for positive semidefinite solutions of the Lyapunov–Stein equation (1.3).

COROLLARY 6.2. *Let $P$ be a positive semidefinite solution of the Lyapunov–Stein equation (1.3) and let (5.12) hold. Then (1.21) holds also.*

PROOF. By Theorem 6.1, the **aBIP**$(M, N, P, C)$ has a solution $S$. Thus, $P_S = P$ and (1.21) now follows from (1.17) by Remark 1.2. □

THEOREM 6.3. *Let $P$ be a positive semidefinite solution of the Lyapunov–Stein equation (1.3), let condition (5.12) be in force and let $\nu$ be the integer defined in (5.5). Then the set $\mathcal{S}(M, N, P, C)$ consists of exactly one element if and only if $\nu = \min(p, q)$; it consists of infinitely many elements if $\nu < \min(p, q)$.*

PROOF. Let $\nu = \min(p, q)$. By Corollary 5.11, the $\widehat{\mathbf{aBIP}}(M, N, P, C)$ has a unique solution $S = \Psi_{12}$ which is defined in (5.19). Since condition (5.12) is in force, the set $\mathcal{S}(M, N, P, C)$ is not empty and hence must consist of the unique element $\Psi_{12}$ due to the inclusion (1.20).

Let $\nu < \min(p, q)$. Then the set $\widehat{\mathcal{S}}(M, N, P, C)$ consists of infinitely many elements and all of them are parametrized by the formula (5.17). Since condition (5.12) is in force, the rational mvf $\Delta_\omega^{[-1]}(z)$ is analytic in $\overline{\mathbb{D}}$ and it follows readily from (5.18)–(5.21) that the coefficients $\Psi_{jk}$ of the Redheffer transform (5.17) are rational mvf's that are analytic in $\overline{\mathbb{D}}$. Let $\mathcal{E}$ be an arbitrary rational function from $\mathcal{S}^{(p-\nu)\times(q-\nu)}$ such that

$$\|\Psi_{21}(z)\mathcal{E}(z)\| < 1 \quad \text{at every point} \quad z \in \overline{\mathbb{D}}. \tag{6.6}$$

We show that the function $S$ corresponding to such a parameter via formula (5.17) is a solution of the **aBIP**$(M, N, P, C)$. To this end set

$$S_\varepsilon(z) = \Psi_{11}(z)\mathcal{E}(z)\left(I_q - \Psi_{21}(z)\mathcal{E}(z)\right)^{-1}\Psi_{22}(z)$$

so that

$$S(z) = \Psi_{12}(z) + S_\varepsilon(z) \tag{6.7}$$

and let $W$ be the mvf constructed via (3.10). Upon substituting (6.7) into (3.10) we get

$$W(z) = W_0(z) - W_\varepsilon(z),$$

where $W_0$ is given by (6.1) and

$$W_\varepsilon(z) = H(z)^{-1}C_1^*S_\varepsilon(z)C_2 G(z)^{-1}.$$

Making use of the equalities (5.29) and (5.30), we get

$$\begin{aligned}W_\varepsilon(z) &= H(z)^{-1}C_1^*\Psi_{11}(z)\mathcal{E}(z)\left(I_q - \Psi_{21}(z)\mathcal{E}(z)\right)^{-1}\Psi_{22}(z)C_2 G(z)^{-1}\\ &= -P^{\frac{1}{2}}\Upsilon_\omega(z)Y_2\mathcal{E}(z)\left(I_q - \Psi_{21}(z)\mathcal{E}(z)\right)^{-1}Y_1^*\Upsilon_\omega(z)P^{\frac{1}{2}},\end{aligned}$$

where $\Upsilon_\omega$ is the mvf defined in (5.31) and $Y_1$ and $Y_2$ are constant matrices from the block decompositions (5.16). Since $\Delta_\omega^{[-1]}(z)$ is analytic in $\overline{\mathbb{D}}$, it follows from (5.31) that the rational mvf $\Upsilon_\omega$ is analytic in $\overline{\mathbb{D}}$ as well. Therefore, the rational mvf $W_\varepsilon$ is analytic in $\overline{\mathbb{D}}$ by (6.6) and $W_0$ is analytic in $\overline{\mathbb{D}}$ by (6.3). Thus, the rational mvf $W$ is analytic in $\overline{\mathbb{D}}$. By Lemma 3.9, $S$ of the form (6.7) is a solution of the **aBIP**$(M, N, P, C)$. This completes the proof, since there are infinitely many parameters $\mathcal{E}$ which are rational and satisfy (6.6). □

COROLLARY 6.4. *Let $P$ be a positive definite solution of the Lyapunov–Stein equation (1.3) and let condition (1.21) be in force. Then the* **aBIP**$(M, N, P, C)$ *has infinitely many solutions.*

PROOF. Since $P > 0$, the number $\nu$ defined by (5.5) is equal to zero, which is certainly less than $\min(p, q)$. In view of Theorem 6.3, it remains only to show that condition (5.12) is in force. Under assumption (1.5), $\operatorname{Ker} M \cap \operatorname{Ker} N = \{0\}$ and, since $P$ is invertible, $\mathcal{K} = \{0\}$, by definition (5.8). Thus, it suffices to show that $\mathcal{K}_\zeta = \{0\}$ for every point $\zeta \in \mathbb{T}$. To this end, let $\zeta_0 \in \mathbb{T}$ and let $f \in \mathcal{K}_{\zeta_0}$. Then, since $P > 0$,

$$Mf = \zeta_0 Nf \quad \text{and} \quad Cf = 0. \tag{6.8}$$

By Remark 2.12, we can assume without loss of generality that the matrices $M$ and $N$ are of the form (2.11). Let

$$f = \operatorname{col}(f_1, f_2, f_3) \quad \text{and} \quad C = \left(\widetilde{C}_1, \widetilde{C}_2, \widetilde{C}_3\right) \tag{6.9}$$

be partitioned conformally with the block decompositions (2.11) of $M$ and $N$. Substituting (2.11) and (6.9) into (6.8) and taking advantage of the spectral conditions (2.10), we conclude that

$$f_1 = 0, \quad f_2 = 0, \quad f_3 = \zeta_0 A_3 f_3 \quad \text{and} \quad \widetilde{C}_3 f_3 = 0.$$

Then

$$\widetilde{C}_3 A_3^k f_3 = \bar{\zeta}_0^{\,k} \widetilde{C}_3 f_3 = 0 \quad (k \geq 0)$$

and therefore,

$$\widetilde{C}_3 \left(I_{k_3} - \zeta A_3\right)^{-1} f_3 = 0$$

at every point $\zeta \in \mathbb{T}$ where $(I_{k_3} - \zeta A_3)$ is invertible. Since the entries $f_1$ and $f_2$ of $f$ are equal to zero, the latter equality implies that

$$C(M - \zeta N)^{-1} f = 0$$

at every point $\zeta \in \mathbb{T}$ where $G(\zeta)$ is invertible. By condition (1.21), $f$ belongs to $\operatorname{Ker} P$ and since $P$ is positive definite, $f = 0$. Thus, $\mathcal{K}_\zeta = \{0\}$ for every point $\zeta \in \mathbb{T}$, as needed. □

## 7. Nontangential limits

In what follows we shall focus on the case when all the singular points of $G^{-1}$ fall on $\mathbb{T}$. The corresponding interpolation problems are then expressed in terms of nontangential limits at these points. In this section we shall prepare a number of basic facts on nontangential limits for mvf's which are analytic in the open unit disk and especially, for functions from the classes $\mathbf{H}_2$ and $\mathcal{C}$. First we review some needed definitions.

For $\beta \in \mathbb{T}$ and $\phi \in (0, \frac{\pi}{2})$, the "ice cream" cone

$$U_\beta(\phi) = \{z : |z| < 1 \text{ and } |\arg(\beta - z)| < \phi\} \tag{7.1}$$

is called *a Stoltz angle* with vertex at $\beta$ and half angle $\phi$ (see e.g. [**44**, Chapter 1]). It is obviously symmetric with respect to the radius connecting $\beta$ with the origin.

Let $\mathcal{U}_\beta \subset \mathbb{D}$ be an open set whose boundary $\partial \mathcal{U}_\beta$ is a rectifiable Jordan curve. We shall say that $\mathcal{U}_\beta$ is a nontangential neighborhood of a point $\beta \in \mathbb{T}$ if it belongs to some Stoltz angle $U_\beta(\phi)$ and has $\beta$ as a limit point.

We say that a mvf $F$ which is analytic in $\mathbb{D}$, has the *nontangential limit* $A$ at $\beta \in \mathbb{T}$ and write $\angle \lim_{z \to \beta} F(z) = A$ if

$$F(z) \to A \quad \text{as} \quad z \to \beta \quad \text{and} \quad z \in U_\beta(\phi) \quad (0 < \phi < \frac{\pi}{2}).$$

It turns out that the boundary behavior of the Carathéodory function $\mathbf{W}$ defined in (3.12), is closely connected to the boundary interpolation conditions imposed on $S$. Therefore, the following lemma will be useful.

LEMMA 7.1. *Let $\mathbf{\Phi}$ belong to $\mathcal{C}^{n \times n}$ and let*

$$\mathbf{\Phi}(z) = i\alpha + \frac{1}{2\pi} \int_0^{2\pi} \frac{e^{it} + z}{e^{it} - z} d\sigma(t) \tag{7.2}$$

*be its Riesz–Herglotz representation with $\alpha = \alpha^*$ and a positive semidefinite $n \times n$-matrix-valued measure $d\sigma(t)$. Let $\beta = e^{it_0} \in \mathbb{T}$ and let $\sigma(\{t_0\})$ denote the matrix assigned to the point $\beta$ by the matrix-valued measure $\sigma$. Then the nontangential limits*

$$\angle \lim_{z \to \beta} (z - \beta)^{j+1} \mathbf{\Phi}^{(j)}(z) = (-1)^{j+1} \frac{j!\beta}{\pi} \sigma(\{t_0\}) \tag{7.3}$$

*and*

$$\angle \lim_{z \to \beta} \left((z - \beta)^{j+1} \mathbf{\Phi}(z)\right)^{(j)} = -\frac{j!\beta}{\pi} \sigma(\{t_0\})$$

*exist for all integers $j \geq 0$.*

PROOF. By assumption, the points $z \in \mathbb{D}$ of interest tend to a boundary point $\beta$ and belong to a Stoltz cone $U_\beta(\phi)$ for some angle $\phi < \frac{\pi}{2}$. Since $z$ tends to $\beta$, it can be assumed without loss of generality that $z$ also belongs to the cone

$$V_\beta(\psi) = \{z : (|\arg z - \arg \beta| < \psi < \frac{\pi}{2} - \phi\} \tag{7.4}$$

with vertex at the origin and half angle $\psi < \frac{\pi}{2} - \phi$. Thus, one can assume $z$ to be of the form

$$z = re^{i\tau}\beta, \quad \text{where} \quad |\arg(z - \beta)| < \phi \quad \text{and} \quad |\tau| < \psi < \frac{\pi}{2} - \phi. \tag{7.5}$$

It follows from (7.5) and the law of sines that $r < \frac{\sin \phi}{\sin(\phi + \tau)}$ and therefore,

$$\left| \frac{\sin \frac{\tau}{2}}{1 - r} \right| < \left| \frac{\sin \frac{\tau}{2}}{1 - \frac{\sin \phi}{\sin(\phi + \tau)}} \right| = \left| \frac{\sin(\phi + \tau)}{2 \cos(\phi + \frac{\tau}{2})} \right| < \frac{1}{2 \cos(\phi + \psi)},$$

which in turn, leads to the estimate

$$\left| \frac{z - \beta}{e^{it} - z} \right| < \frac{|z - \beta|}{1 - |z|} = \frac{|1 - re^{i\tau}|}{1 - r}$$

$$= \left| 1 + r \frac{1 - e^{i\tau}}{1 - r} \right| < 1 + 2 \left| \frac{\sin \frac{\tau}{2}}{1 - r} \right| < 1 + \frac{1}{\cos(\phi + \psi)}. \tag{7.6}$$

It follows from (7.2) that

$$\mathbf{\Phi}^{(j)}(z) = \frac{j!}{\pi} \int_0^{2\pi} \frac{e^{it}}{(e^{it} - z)^{j+1}} d\sigma(t) \quad (j \geq 1)$$

and therefore,

$$(z - \beta)^{j+1} \mathbf{\Phi}^{(j)}(z) = \frac{j!}{\pi} \int_0^{2\pi} f_z(t) d\sigma(t),$$

where
$$f_z(t) = \frac{(z-\beta)^{j+1}e^{it}}{(e^{it}-z)^{j+1}}.$$

By (7.6), $f_z$ has a summable majorant and it is readily seen that
$$\angle\lim_{z\to\beta} f_z(t) = \begin{cases} 0 & \text{if } e^{it} \neq \beta, \\ (-1)^{j+1}\beta & \text{if } e^{it} = \beta. \end{cases}$$

Therefore, by the dominated convergence principle,
$$\angle\lim_{z\to\beta}(z-\beta)^{j+1}\boldsymbol{\Phi}^{(j)}(z) = \frac{j!}{\pi}\int_0^{2\pi}\angle\lim_{z\to\beta} f_z(t)d\sigma(t)$$
$$= (-1)^{j+1}\frac{j!\beta}{\pi}\sigma(\{t_0\}) \quad (e^{it_0} = \beta),$$

which proves (7.3) for $j \geq 1$. The case $j = 0$ is established in much the same way with the help of the representation (7.2) (see e.g., [**23**, Lemma 8.1]).

The second assertion of the lemma follows from (7.3): by Leibnitz's rule,
$$\angle\lim_{z\to\beta}\left((z-\beta)^{k+1}\boldsymbol{\Phi}(z)\right)^{(k)} = \angle\lim_{z\to\beta}\sum_{j=0}^{k}\binom{k}{j}\frac{(k+1)!}{(j+1)!}(z-\beta)^{j+1}\boldsymbol{\Phi}^{(j)}(z)$$
$$= \sum_{j=0}^{k}\binom{k+1}{j+1}\frac{k!}{j!}\angle\lim_{z\to\beta}(z-\beta)^{j+1}\boldsymbol{\Phi}^{(j)}(z)$$
$$= \frac{k!\beta}{\pi}\sigma(\{t_0\})\sum_{j=0}^{k}(-1)^{j+1}\binom{k+1}{j+1}$$
$$= -\frac{k!\beta}{\pi}\sigma(\{t_0\}).$$

□

LEMMA 7.2. *Let $\boldsymbol{\Phi}$ belong to $\mathcal{C}^{n\times n}$ and let*
$$\Phi(z) = \frac{\boldsymbol{\Phi}(z) - \boldsymbol{\Phi}(0)}{z}.$$

*Then the nontangential limits*
$$\angle\lim_{z\to\beta}(z-\beta)^{j+1}\Phi(z)^{(j)} = (-1)^{j+1}\frac{j!}{\pi}\sigma(\{t_0\})$$
*and*
$$\angle\lim_{z\to\beta}\left((z-\beta)^{j+1}\Phi(z)\right)^{(j)} = -\frac{j!}{\pi}\sigma(\{t_0\})$$
*exist for all $j \geq 0$, where $\beta = e^{it_0} \in \mathbb{T}$ and $\sigma(\{t_0\})$ is the measure assigned to the point $t_0$ by the matrix-valued measure $\sigma$ from the Riesz–Herglotz representation (7.2) of $\boldsymbol{\Phi}$.*

PROOF. By (7.2),
$$\boldsymbol{\Phi}(z) = \frac{1}{2z\pi}\int_0^{2\pi}\left(\frac{e^{it}+z}{e^{it}-z}-1\right)d\sigma(t) = \frac{1}{\pi}\int_0^{2\pi}\frac{d\sigma(t)}{e^{it}-z}$$

and
$$\Phi^{(j)}(z) = \frac{j!}{\pi}\int_0^{2\pi}\frac{d\sigma(t)}{(e^{it}-z)^{j+1}}.$$

The rest of the proof is much the same as the proof of Lemma 7.1. □

The Riesz–Herglotz representation (7.2) of $\mathbf{\Phi}$ guarantees that the kernel

$$\mathbf{K_\Phi}(z,\omega) = \frac{\mathbf{\Phi}(z) + \mathbf{\Phi}(\omega)^*}{1 - z\bar{\omega}} \tag{7.7}$$

admits the integral representation

$$\mathbf{K_\Phi}(z,\omega) = \frac{1}{\pi}\int_0^{2\pi} \frac{d\sigma(t)}{(e^{it}-z)(e^{-it}-\bar{\omega})},$$

and hence is a positive kernel on $\mathbb{D} \times \mathbb{D}$. Moreover, since

$$\frac{\partial^{j+\ell}}{\partial z^j \partial \bar{\omega}^\ell}\left(\mathbf{K_\Phi}(z,\omega)\right) = \frac{j!\ell!}{\pi}\int_0^{2\pi} \frac{d\sigma(t)}{(e^{it}-z)^{j+1}(e^{-it}-\bar{\omega})^{\ell+1}}, \tag{7.8}$$

the arguments used to prove Lemma 7.1 lead in much the same way to the following two conclusions:

LEMMA 7.3. *Let $\mathbf{\Phi} \in \mathcal{C}^{n \times n}$ admit a representation (7.2) with a positive semi-definite matrix-valued measure $d\sigma$, let $\mathbf{K_\Phi}$ be the associated positive kernel defined by (7.7) and let $\beta = e^{it_0} \in \mathbb{T}$. Then the nontangential limits*

$$\angle \lim_{z,\omega \to \beta} (z-\beta)^{j+1}(\bar{\omega}-\bar{\beta})^{\ell+1} \frac{\partial^{j+\ell}}{\partial z^j \partial \bar{\omega}^\ell}\left(\mathbf{K_\Phi}(z,\omega)\right) = (-1)^{j+\ell}\frac{j!\ell!}{\pi}\sigma(\{t_0\}) \tag{7.9}$$

*and*

$$\angle \lim_{z,\omega \to \beta} \frac{\partial^{j+\ell}}{\partial z^j \partial \bar{\omega}^\ell}\left((z-\beta)^{j+1}(\bar{\omega}-\bar{\beta})^{\ell+1}\mathbf{K_\Phi}(z,\omega)\right) = \frac{j!\ell!}{\pi}\sigma(\{t_0\})$$

*exist for every pair of nonnegative integers $j$ and $\ell$.*

PROOF. The proof of the first assertion relies on the integral representation (7.8) and the arguments from the proof of Lemma 7.1. The second assertion follows from (7.9) by a double application of Leibnitz's rule:

$$\angle \lim_{z,\omega \to \beta} \frac{\partial^{j+\ell}}{\partial z^j \partial \bar{\omega}^\ell}\left((z-\beta)^{j+1}(\bar{\omega}-\bar{\beta})^{\ell+1}\mathbf{K_\Phi}(z,\omega)\right)$$

$$= \angle \lim_{z,\omega \to \beta} \sum_{i=0}^{j}\sum_{k=0}^{\ell} \binom{j}{i}\binom{\ell}{k} \frac{(j+1)!(\ell+1)!}{(i+1)!(k+1)!}$$

$$\times (z-\beta)^{i+1}(\bar{\omega}-\bar{\beta})^{k+1}\frac{\partial^{i+k}}{\partial z^i \partial \bar{\omega}^k}\left(\mathbf{K_\Phi}(z,\omega)\right)$$

$$= \frac{j!\ell!}{\pi}\sigma(\{t_0\})\sum_{i=0}^{j}\sum_{k=0}^{\ell}(-1)^{i+k}\binom{j+1}{i+1}\binom{\ell+1}{k+1} = \frac{j!\ell!}{\pi}\sigma(\{t_0\}).$$

□

LEMMA 7.4. *Let $\mathbf{\Phi} \in \mathcal{C}^{n \times n}$, let $\mathbf{K_\Phi}$ be the associated positive kernel defined by (7.7) and let $\beta_1$ and $\beta_2$ be two distinct points on $\mathbb{T}$. Then*

$$\angle \lim_{\substack{z \to \beta_1 \\ \omega \to \beta_2}} \frac{\partial^{j+\ell}}{\partial z^j \partial \bar{\omega}^\ell}\left((z-\beta_1)^{j+1}(\bar{\omega}-\bar{\beta}_2)^{\ell+1}\mathbf{K_\Phi}(z,\omega)\right)$$

$$= \angle \lim_{\substack{z \to \beta_1 \\ \omega \to \beta_2}} (z-\beta_1)^{j+1}(\bar{\omega}-\bar{\beta}_2)^{\ell+1}\frac{\partial^{j+\ell}}{\partial z^j \partial \bar{\omega}^\ell}\left(\mathbf{K_\Phi}(z,\omega)\right) = 0 \tag{7.10}$$

*for every pair of nonnegative integers $j$ and $\ell$.*

PROOF. Using (7.8) we get

$$(z-\beta_1)^{j+1}(\bar{\omega}-\bar{\beta}_2)^{\ell+1}\frac{\partial^{j+\ell}}{\partial z^j \partial \bar{\omega}^\ell}(\mathbf{K}_\Phi(z,\omega)) = \frac{j!\ell!}{\pi}\int_0^{2\pi} f_{z,\omega}(t)d\sigma(t),$$

where

$$f_{z,\omega}(t) = \frac{(z-\beta_1)^{j+1}(\bar{\omega}-\bar{\beta}_2)^{\ell+1}}{(e^{it}-z)^{j+1}(e^{-it}-\bar{\omega})^{\ell+1}}.$$

It is readily seen that

$$\angle \lim_{\substack{z\to\beta_1\\ \omega\to\beta_2}} f_{z,\omega}(t) = 0 \quad (\forall\, t)$$

and then, the second equality in (7.10) follows by the dominated convergence principle (which is applicable, thanks to (7.6)). The first assertion now follows just as in the proof of Lemma 7.3, by a double application of Leibnitz's rule. □

LEMMA 7.5. *Let $F \in \mathbf{H}_2^{p\times q}$. Then, for every integer $j \geq 0$ there exists a constant $A_j > 0$ such that*

$$\left\|F^{(j)}(z)\right\| \leq \frac{A_j\|F\|_{\mathbf{H}_2^{p\times q}}}{(1-|z|^2)^{j+\frac{1}{2}}} \tag{7.11}$$

*for every $z \in \mathbb{D}$.*

PROOF. Let $\rho_z$ be defined as in (2.4) and let $y \in \mathbb{C}^p$. Then, since

$$\frac{\partial^j}{\partial \bar{z}^j}\left(\frac{I_p}{\rho_z(\zeta)}\right) = j!\frac{\zeta^j I_p}{\rho_z(\zeta)^{j+1}}$$

is the reproducing kernel for the $j$-th derivative in $\mathbf{H}_2^p$,

$$\left\|\frac{\zeta^j y}{\rho_z(\zeta)^{j+1}}\right\|^2_{\mathbf{H}_2^p} = \left\langle \frac{\zeta^j y}{\rho_z(\zeta)^{j+1}}, \frac{\zeta^j y}{\rho_z(\zeta)^{j+1}}\right\rangle_{\mathbf{H}_2^p}$$

$$= \frac{y^*y}{j!}\left(\frac{\zeta^j}{\rho_z(\zeta)^{j+1}}\right)^{(j)}\bigg|_{\zeta=z}$$

$$= \|y\|^2 \sum_{k=0}^{j}\frac{(j+k)!}{(k!)^2(j-k)!}\cdot\frac{|z|^{2k}}{(1-|z|^2)^{j+k+1}}$$

$$\leq \frac{\|y\|^2}{\rho_z(z)^{2j+1}}\sum_{k=0}^{j}\frac{(j+k)!}{(k!)^2(j-k)!}.$$

Using again the reproducing kernel property together with Cauchy's inequality and setting

$$A_j = j!\left(\sum_{k=0}^{j}\frac{(j+k)!}{(k!)^2(j-k)!}\right)^{\frac{1}{2}},$$

we conclude that for every choice of $x \in \mathbb{C}^q$,

$$|y^* F^{(j)}(z)x| = j! \left| \left\langle Fx, \frac{\zeta^j y}{\rho_z(\zeta)^{j+1}} \right\rangle_{\mathbf{H}_2^p} \right| \leq j! \, \|Fx\|_{\mathbf{H}_2^p} \cdot \left\| \frac{\zeta^j y}{\rho_z(\zeta)^{j+1}} \right\|_{\mathbf{H}_2^p}$$

$$\leq \frac{A_j \cdot \|F\|_{\mathbf{H}_2^{p \times q}} \cdot \|x\| \cdot \|y\|}{(1-|z|^2)^{j+\frac{1}{2}}}.$$

Choosing $y = F^{(j)}(z)x$ in the preceding inequality, we get

$$\|F^{(j)}(z)x\|^2 \leq \frac{A_j \cdot \|F\|_{\mathbf{H}_2^{p \times q}} \cdot \|x\| \cdot \|F^{(j)}(z)x\|}{(1-|z|^2)^{j+\frac{1}{2}}},$$

which is equivalent to

$$\|F^{(j)}(z)x\| \leq \frac{A_j \cdot \|F\|_{\mathbf{H}_2^{p \times q}} \|x\|}{(1-|z|^2)^{j+\frac{1}{2}}}$$

and leads easily to (7.11). □

COROLLARY 7.6. *Let* $F \in \mathbf{H}_2^{p \times q}$ *and let* $\beta \in \mathbb{T}$. *Then, for every pair of integers $j \geq 0$ and $k \geq 1$, the nontangential limit*

$$\angle \lim_{z \to \beta} \frac{d^j}{dz^j} \left( (z-\beta)^{j+k} F(z) \right) = 0$$

*exists.*

PROOF. Let $z$ be of the form (7.5). Then it follows from (7.11) and (7.6) that

$$\|(z-\beta)^{j+k} F^{(j)}(z)\| \leq A_j \|F\|_{\mathbf{H}_2^{p \times q}} \left| \frac{1-re^{i\tau}}{1-r^2} \right|^{j+\frac{1}{2}} |z-\beta|^{k-\frac{1}{2}}$$

$$< A_j \|F\|_{\mathbf{H}_2^{p \times q}} \left( 1 + \frac{1}{\cos(\phi+\psi)} \right)^{j+\frac{1}{2}} |z-\beta|^{k-\frac{1}{2}}$$

and therefore, that

$$\angle \lim_{z \to \beta} (z-\beta)^{j+k} F^{(j)}(z) = 0 \qquad (j \geq 0, \, k \geq 1).$$

The rest follows easily by Leibnitz's rule. □

The next lemma is an analogue of Lemma 7.3 for Hardy functions.

LEMMA 7.7. *Let* $F \in \mathbf{H}_2^{n \times q}$ *and let* $\beta \in \mathbb{T}$. *Then, for every nonnegative integer $m$, the nontangential limit*

$$\angle \lim_{z, \omega \to \beta} \frac{\partial^{2m}}{\partial z^m \partial \bar{\omega}^m} \left( (z-\beta)^{m+1} \frac{F(z)F(\omega)^*}{\rho_\omega(z)} (\bar\omega - \bar\beta)^{m+1} \right) = 0. \qquad (7.12)$$

PROOF. Let

$$\Phi(z) = \frac{1}{4\pi} \int_0^{2\pi} \frac{e^{it}+z}{e^{it}-z} F(e^{it}) F(e^{it})^* dt \qquad (7.13)$$

so that

$$\frac{\Phi(z)+\Phi(\omega)^*}{1-z\bar\omega} = \frac{1}{2\pi} \int_0^{2\pi} \frac{F(e^{it})F(e^{it})^* dt}{(e^{it}-z)(e^{-it}-\bar\omega)}.$$

The last equality implies that

$$v^* \frac{\Phi(z) + \Phi(\omega)^*}{1 - z\bar{\omega}} u = \left\langle \frac{F^*u}{\rho_\omega}, \frac{F^*v}{\rho_z} \right\rangle_{L_2^p}^2$$

for $z, \omega \in \mathbb{D}$ and $u, v \in \mathbb{C}^n$.

Let $\underline{p}$ denote the orthogonal projection of $L_2^p(\mathbb{T})$ onto $\mathbf{H}_2^p$. Then, since $F \in \mathbf{H}_2^{n \times q}$,

$$\underline{p} \frac{F^*u}{\rho_z} = \frac{F(z)^*u}{\rho_z}$$

for $z \in \mathbb{D}$ and $u \in \mathbb{C}^n$ (for a proof see e.g., [23, Lemma 2.1]), it is readily seen that for every set of points $\omega_1, \ldots, \omega_m \in \mathbb{D}$, of vectors $u_1, \ldots, u_m \in \mathbb{C}^n$ and every positive integer $m$,

$$\sum_{i,j=1}^m u_i^* \frac{F(\omega_i)F(\omega_j)^*}{\rho_{\omega_j}(\omega_i)} u_j = \left\| \sum_{j=1}^m \frac{F(\omega_j)^*u_j}{\rho_{\omega_j}} \right\|_{\mathbf{H}_2^p}^2 = \left\| \underline{p} \sum_{j=1}^m \frac{F^*u_j}{\rho_{\omega_j}} \right\|_{L_2^p}^2$$

$$\leq \left\| \sum_{j=1}^m \frac{F^*u_j}{\rho_{\omega_j}} \right\|_{L_2^p}^2 = \sum_{i,j=1}^m u_i^* \frac{\Phi(\omega_i) + \Phi(\omega_j)^*}{\rho_{\omega_j}(\omega_i)} u_j.$$

This means that

$$\frac{\Phi(z) + \Phi(\omega)^* - F(z)F(\omega)^*}{\rho_\omega(z)} \succeq 0 \quad (z, \omega \in \mathbb{D}). \tag{7.14}$$

Therefore, the kernels

$$K_\omega^1(z) = \frac{\partial^{2m}}{\partial z^m \partial \bar{\omega}^m} \left( (z - \beta)^{m+1} \frac{\Phi(z) + \Phi(\omega)^*}{\rho_\omega(z)} (\bar{\omega} - \bar{\beta})^{m+1} \right)$$

and

$$K_\omega^2(z) = \frac{\partial^{2m}}{\partial z^m \partial \bar{\omega}^m} \left( (z - \beta)^{m+1} \frac{F(z)F(\omega)^*}{\rho_\omega(z)} (\bar{\omega} - \bar{\beta})^{m+1} \right),$$

are positive on $\mathbb{D} \times \mathbb{D}$, by Proposition 2.3, and satisfy

$$K_\omega^1(z) \succeq K_\omega^2(z) \succeq 0,$$

on account of (7.14). By the proof of Proposition 2.2,

$$|x^* K_\omega^2(z) y|^2 \leq (x^* K_z^2(z) x)(y^* K_\omega^2(\omega) y) \leq (x^* K_z^1(z) x)(y^* K_\omega^1(\omega) y) \tag{7.15}$$

for every $x, y \in \mathbb{C}^n$. Since the measure in the Riesz–Herglotz representation (7.13) of $\Phi$ is absolutely continuous, it follows by Lemma 7.3 that

$$\angle \lim_{z,\omega \to \beta} K_\omega^1(z) = 0,$$

which together with (7.15) implies that

$$\angle \lim_{z,\omega \to \beta} x^* K_\omega^2(z) y = 0.$$

The latter is equivalent to (7.12), since $x$ and $y$ are arbitrary. □

## 7. NONTANGENTIAL LIMITS

LEMMA 7.8. *Let $\beta \in \mathbb{T}$ and let $F$ be a mvf which is analytic in $\mathbb{D}$. Then*

$$\angle \lim_{z \to \beta} (z - \beta)^{-m} F(z) = 0 \tag{7.16}$$

*if and only if*

$$\angle \lim_{z \to \beta} F^{(j)}(z) = 0 \quad \text{for} \quad j = 0, \ldots, m. \tag{7.17}$$

PROOF. Let relation (7.16) hold. By the definition of nontangential limit, it suffices to show that $F^{(j)}(z)$ tends to zero as $z$ tends to $\beta$ from inside any preassigned Stoltz angle $U_\beta(\phi)$ defined via (7.1). Take $\widetilde{\phi}$ ($\phi < \widetilde{\phi} < \frac{\pi}{2}$) and a Stoltz angle $U_\beta(\widetilde{\phi}) \supset U_\beta(\phi)$. By (7.16), for every $\varepsilon > 0$, there exists $\delta > 0$ such that

$$\|(\zeta - \beta)^{-j} F(\zeta)\| < \varepsilon \quad \text{for} \quad j = 0, \ldots, m \text{ and } \forall \zeta \in U_\beta(\widetilde{\phi}) \cap \mathbb{D}_\beta(\delta), \tag{7.18}$$

where $\mathbb{D}_\beta(\delta) = \{z : |z - \beta| < \delta\}$. Take $z \in U_\beta(\phi) \cap \mathbb{D}_\beta(\delta/2)$ and let $\mathbb{T}_z$ be the maximal circle which is centered at $z$ and lies in $U_\beta(\widetilde{\phi})$. It is easily seen that $\mathbb{T}_z$ also belongs to $\mathbb{D}_\beta(\delta)$. Therefore, the estimate (7.18) is valid for every point $\zeta \in \mathbb{T}_z$. Moreover, the inequality

$$\left|\frac{\zeta - \beta}{\zeta - z}\right| < 1 + \left|\frac{z - \beta}{\zeta - z}\right| < 1 + \frac{1}{\sin(\widetilde{\phi} - \phi)} \tag{7.19}$$

holds for all $\zeta \in \mathbb{T}_z$ and thus, by Cauchy's formula,

$$\begin{aligned}
\|F^{(j)}(z)\| &= \frac{j!}{2\pi} \left\| \int_{\mathbb{T}_z} \frac{F(\zeta) d\zeta}{(\zeta - z)^{j+1}} \right\| \\
&\leq \frac{j!}{2\pi} \left\| \int_{\mathbb{T}_z} \frac{d\zeta}{\zeta - z} \right\| \max_{\zeta \in \mathbb{T}_z} \left\{ \frac{\|F(\zeta)\|}{|\zeta - \beta|^j} \left|\frac{\zeta - \beta}{\zeta - z}\right|^j \right\} \\
&< j! \varepsilon \left(1 + \frac{1}{\sin(\widetilde{\phi} - \phi)}\right)^j,
\end{aligned}$$

which proves (7.17).

Conversely, if the relations (7.17) are in force, take $\varepsilon > 0$ and choose $\delta > 0$ such that

$$\|F^{(m)}(z)\| < \varepsilon \quad \text{for all} \quad z \in U_\beta(\phi) \cap \mathbb{D}_\beta(\delta). \tag{7.20}$$

Then, upon letting $\omega \to \beta$ in the Taylor representation

$$F(z) = \sum_{j=0}^{m-1} \frac{F^{(j)}(\omega)}{j!} (z - \omega)^j + \int_\omega^z \frac{(z - \zeta)^{m-1}}{(m-1)!} F^{(m)}(\zeta) d\zeta \quad (\omega \in U_\beta(\phi) \cap \mathbb{D}_\beta(\delta)),$$

and taking advantage of (7.17), we get

$$F(z) = \frac{1}{(m-1)!} \int_\beta^z (z - \zeta)^{m-1} F^{(m)}(\zeta) d\zeta.$$

Using the estimate (7.20) and the parametrization $\zeta = \beta + (z - \beta)t$ ($0 \leq t \leq 1$), we get

$$\|F(z)\| \leq \frac{\varepsilon}{(m-1)!} \int_0^1 |z - \beta|^m (1-t)^{m-1} dt = \frac{\varepsilon}{m!} |z - \beta|^m,$$

which proves (7.16). □

COROLLARY 7.9. *Let $\beta \in \mathbb{T}$ and let $F$ be a mvf which is analytic in $\mathbb{D}$. Then*

$$\angle \lim_{z \to \beta} (z-\beta)^{-m} \{F(z) - F_0 - (z-\beta)F_1 - \ldots - F_m(z-\beta)^m\} = 0, \qquad (7.21)$$

*if and only if*

$$\angle \lim_{z \to \beta} \frac{F^{(j)}(z)}{j!} = F_j \quad \text{for} \quad j = 0, \ldots, m. \qquad (7.22)$$

For the proof it suffices to apply the preceding lemma to the function

$$\widetilde{F}(z) = F(z) - F_0 - (z-\beta)F_1 - \ldots - F_m(z-\beta)^m.$$

REMARK 7.10. Let $\mathcal{U} \subset \mathbb{D}$ be any simply connected open set whose boundary is a rectifiable Jordan curve and let $F$ be a mvf which is analytic in $\mathbb{D}$ and is such that $F'$ is uniformly bounded on $\mathcal{U}$:

$$\|F'(z)\| \leq k < \infty \quad (z \in \mathcal{U} \subseteq \mathbb{D}).$$

Then $F$ is uniformly bounded on $\mathcal{U}$.

For the proof we fix a point $\alpha \in \mathcal{U}$, represent $F$ in the form

$$F(z) = F(\alpha) + \int_\alpha^z F'(\zeta) d\zeta$$

and conclude that

$$\|F(z)\| \leq \|F(\alpha)\| + L(\partial \mathcal{U})\|F'(z)\| < \|F(\alpha)\| + kL(\partial \mathcal{U}) \quad (z \in \mathcal{U}),$$

where $L(\partial \mathcal{U})$ denotes the length of the rectifiable curve $\partial \mathcal{U}$.

COROLLARY 7.11. *Let a kernel $K(z, \omega)$ be analytic in $z$ and $\bar{\omega}$ on $\mathbb{D}$ let $\dfrac{\partial^{2m} K}{\partial z^m \partial \bar{\omega}^m}$ be uniformly bounded for every pair of points $z$ and $\omega$ in a nontangential neighborhood $\mathcal{U}_\beta$ of $\beta \in \mathbb{T}$. Then*

$$\widetilde{K}(z, \omega) = \frac{\partial^{2m}}{\partial z^m \partial \bar{\omega}^m} \left( (z-\beta)(\bar{\omega} - \bar{\beta}) K(z, \omega) \right)$$

*is uniformly bounded for every pair of points $z$ and $\omega$ in $\mathcal{U}_\beta$.*

PROOF. By Leibnitz's rule,

$$\begin{aligned}\widetilde{K} &= (z-\beta)(\bar{\omega}-\bar{\beta})\frac{\partial^{2m} K}{\partial z^m \partial \bar{\omega}^m} + m(z-\beta)\frac{\partial^{2m-1} K}{\partial z^m \partial \bar{\omega}^{m-1}} \\ &\quad + m(\bar{\omega} - \bar{\beta})\frac{\partial^{2m-1} K}{\partial z^{m-1} \partial \bar{\omega}^m} + m^2 \frac{\partial^{2m-2} K}{\partial z^{m-1} \partial \bar{\omega}^{m-1}}.\end{aligned}$$

The first term on the right is uniformly bounded by assumption. The three other terms are uniformly bounded by Remark 7.10. □

## 8. The Nevanlinna–Pick boundary problem

In this section we apply the preceding analysis to study a boundary Nevanlinna–Pick problem that will be formulated below in terms of the integers $r_1, \ldots, r_m$, $1 \leq r_j \leq \min(p, q)$, $n = r_1 + \ldots + r_m$ and the entries in the matrices

$$M = I_n, \quad N = \begin{pmatrix} \bar{\beta}_1 I_{r_1} & & \\ & \ddots & \\ & & \bar{\beta}_m I_{r_m} \end{pmatrix}, \quad C = \begin{pmatrix} C_1 \\ C_2 \end{pmatrix} = \begin{pmatrix} \xi_1 & \cdots & \xi_m \\ \eta_1 & \cdots & \eta_m \end{pmatrix}, \tag{8.1}$$

where the $\beta_j$ are distinct points on $\mathbb{T}$, $\xi_j \in \mathbb{C}^{p \times r_j}$ and $\eta_j \in \mathbb{C}^{q \times r_j}$. Correspondingly, by (1.4) and (1.12),

$$G(z)^{-1} = \operatorname{diag}\left\{ \frac{I_{r_1}}{1 - \bar{\beta}_1 z}, \ldots, \frac{I_{r_m}}{1 - \bar{\beta}_m z} \right\} \tag{8.2}$$

and

$$H(z)^{-1} = \operatorname{diag}\left\{ \frac{I_{r_1}}{z - \beta_1}, \ldots, \frac{I_{r_m}}{z - \beta_m} \right\}. \tag{8.3}$$

We shall consider the $\widehat{\mathbf{aBIP}}(I_n, N, P, C)$ for this choice of $N$ and $C$ and for any $n \times n$–matrix $P \geq 0$ satisfying the Stein equation

$$P - N^* P N = C_1^* C_1 - C_2^* C_2. \tag{8.4}$$

We shall assume that

$$P = (P_{j\ell})_{j,\ell=1}^m, \quad P_{j\ell} \in \mathbb{C}^{r_j \times r_\ell} \tag{8.5}$$

is partitioned conformally with $N$. Then, in view of (8.1), the Stein equation (8.4) reduces to the following set of equalities:

$$(1 - \beta_j \bar{\beta}_\ell) P_{j\ell} = \xi_j^* \xi_\ell - \eta_j^* \eta_\ell, \quad j, \ell = 1, \ldots, m. \tag{8.6}$$

The off diagonal blocks $P_{j\ell}$ of $P$ are uniquely defined by (8.6) and are equal to

$$P_{j\ell} = \frac{\xi_j^* \xi_\ell - \eta_j^* \eta_\ell}{1 - \beta_j \bar{\beta}_\ell} \quad (j \neq \ell).$$

On the other hand, the diagonal blocks $P_{jj}$ are not defined by (8.6) but the conditions

$$\xi_j^* \xi_j = \eta_j^* \eta_j \quad \text{for} \quad j = 1, \ldots, m \tag{8.7}$$

are seen to be necessary for the Stein equation (8.4) to have a solution. They are also sufficient to insure the existence of a positive semidefinite solution $P \geq 0$: if (8.7) is in force, we define the off diagonal blocks of $P$ by (8.5) and then choose the diagonal blocks $P_{jj}$ of $P$ large enough to insure the positivity of $P$.

The next theorem, which is established in [**23**, Section 8], serves to express condition (1.9) in the present setting in terms of the boundary limits of an interpolant $S$. The existence and identification of the limits in (8.9) and (8.12) under condition (8.8) was obtained earlier in [**39**] by different methods.

THEOREM 8.1. *Let $S \in \mathcal{S}^{p \times q}$, $\beta \in \mathbb{T}$ and let $\xi \in \mathbb{C}^{p \times r}$ be a matrix of full rank $r$ ($r \leq \min(p, q)$). Then:*

*I. The following three statements are equivalent:*

(1) The reproducing kernel $\Lambda_\omega(z)$ of $\mathcal{H}(S)$ defined in (2.5), is subject to the bound
$$\|\xi^*\Lambda_z(z)\xi\| \leq \kappa\|\xi\|^2 < \infty \tag{8.8}$$
for every point $z$ in a nontangential neighborhood $\mathcal{U}_\beta$ of $\beta$.

(2) The nontangential limit
$$P_{\mathbf{L}} = \angle \lim_{z\to\beta} \xi^*\Lambda_z(z)\xi \tag{8.9}$$
exists.

(3) The nontangential limit
$$\angle \lim_{z\to\beta} S(z)^*\xi = \eta \tag{8.10}$$
exists and serves to define the matrix $\eta$. Furthermore,
$$\angle \lim_{z\to\beta} S(z)\eta = \xi, \quad \xi^*\xi = \eta^*\eta, \tag{8.11}$$
and the nontangential limit
$$P_{\mathbf{V}} = \angle \lim_{z\to\beta} \frac{\xi^*\xi - \xi^*S(z)\eta}{1 - z\bar\beta} \tag{8.12}$$
exists.

II. If any one (and hence every one) of the preceding three statements is in force, then the columns of the mvf $B(z) = \dfrac{\xi - S(z)\eta}{\rho_\beta(z)}$ belong to $\mathcal{H}(S)$, the nontangential limits (8.12) and (8.9) are equal and
$$x^*P_{\mathbf{L}}x = x^*P_{\mathbf{V}}x = \|Bx\|^2_{\mathcal{H}(S)}$$
for every $x \in \mathbb{C}^n$.

III. Any two of the three equalities in (8.10) and (8.11) imply the third.

Theorem 8.1 is a matrix valued extension of the classical Carathéodory–Julia theorem [19], [32], which may be obtained by setting $p = q = 1$. Additional sources and discussion regarding this classical result can be found in [49, Chapter VI]; see also [20] and [51, Chapter 4] for various applications. An extension of Theorem 8.1 that incorporates high order directional boundary derivatives of a matrix valued Schur function $S$ will be presented in Section 9 (see Theorem 9.1).

Now we formulate the Nevanlinna–Pick boundary problem $\widehat{\mathbf{NPBP}}$ in terms of boundary limits as in [14] and [39].

The $\widehat{\mathbf{NPBP}}$: Given $m$ distinct points $\beta_1,\ldots,\beta_m$ on $\mathbb{T}$ and given matrices $\xi_j \in \mathbb{C}^{p\times r_j}$, $\eta_j \in \mathbb{C}^{q\times r_j}$ and $\gamma_j \in \mathbb{C}^{r_j\times r_j}$ with $\operatorname{rank}\xi_j = r_j$, find necessary and sufficient conditions which insure the existence of a Schur function $S \in \mathcal{S}^{p\times q}$ such that:
$$\angle \lim_{z\to\beta_j} S(z)^*\xi_j = \eta_j \quad \text{and} \quad \angle \lim_{z\to\beta_j} \xi_j^*\frac{I - S(z)S(z)^*}{\rho_z(z)}\xi_j \leq \gamma_j \quad (j=1,\ldots,m). \tag{8.13}$$

The next theorem shows that the $\widehat{\mathbf{NPBP}}$ actually is equivalent to the $\widehat{\mathbf{aBIP}}(I_n, N, P, C)$ with $N$ and $C$ defined by (8.1) for a special choice of $P$ and establishes a solvability criteria for the $\widehat{\mathbf{NPBP}}$.

8. THE NEVANLINNA–PICK BOUNDARY PROBLEM

THEOREM 8.2. *Let $N$ and $C$ be defined as in* (8.1) *and let*

$$P = (P_{j\ell})_{j,\ell=1}^m, \quad \text{where} \quad P_{j\ell} = \begin{cases} \dfrac{\xi_j^* \xi_\ell - \eta_j^* \eta_\ell}{1 - \beta_j \bar{\beta}_\ell} & j \neq \ell \\ \gamma_j & j = \ell. \end{cases} \quad (8.14)$$

*Then:*

(1) *$P$ is a solution of the Stein equation* (8.4) *if and only if* (8.7) *holds.*
(2) *If $S$ is a solution of the $\widehat{\text{NPBP}}$, then $P$ is a positive semidefinite solution of the Stein equation* (8.4) *and $S$ is a solution of the $\widehat{\text{aBIP}}(I_n, N, P, C)$.*
(3) *If $P$ is a positive semidefinite solution of the Stein equation* (8.4) *and $S$ is a solution of the $\widehat{\text{aBIP}}(I_n, N, P, C)$, then $S$ is a solution of the $\widehat{\text{NPBP}}$.*
(4) *The $\widehat{\text{NPBP}}$ has a solution if and only if $P$ is a positive semidefinite solution of the Stein equation* (8.4), *i.e., if and only if* (8.7) *holds and $P \geq 0$. Moreover, if in this case* $\operatorname{rank} P \geq 1$, *then all the solutions of the $\widehat{\text{NPBP}}$ are parametrized by the linear fractional transformation* (5.3) *where $\Theta$ is specified by* (4.24) *(or by the Redheffer transform* (5.17) *that is specified by Theorem 5.9).*

PROOF. The first assertion is readily checked by direct calculation. Suppose next that $S$ is a solution of the $\widehat{\text{NPBP}}$. Then, in particular, the functions $\|\xi_j^* \Lambda_z(z) \xi_j\|$ are uniformly bounded in a nontangential neighborhood of $\beta_j$ for $j = 1, \ldots, m$. Therefore, (8.7) is in force by Theorem 8.1 and hence, by the first assertion, $P$ is a solution of the Stein equation (8.4). By (2.7),

$$\xi_j^* \frac{I_p - S(z)S(z)^*}{\rho_z(z)} \xi_j = [\Psi_{z,j}, \Psi_{z,j}]_S, \quad \text{where} \quad \Psi_{z,j}(\zeta) = \begin{pmatrix} I_p \\ S(z)^* \end{pmatrix} \frac{\xi_j}{\rho_z(\zeta)}.$$

Substituting the decompositions of $C$ and $G(z)$ from (8.1) and (8.2) into (1.10), we see that the block entries of $P_S$ are equal to

$$[P_S]_{j\ell} = \left[ \begin{pmatrix} \xi_\ell \\ \eta_\ell \end{pmatrix} \frac{1}{\rho_{\beta_\ell}}, \begin{pmatrix} \xi_j \\ \eta_j \end{pmatrix} \frac{1}{\rho_{\beta_j}} \right]_S. \quad (8.15)$$

Taking advantage of (8.13)–(8.15) we conclude by Fatou's lemma that

$$\begin{aligned}
[P_S]_{jj} &= \left[ \angle \lim_{z \to \beta_j} \Psi_{z,j}, \angle \lim_{z \to \beta_j} \Psi_{z,j} \right]_S \\
&= \frac{1}{2\pi} \int_0^{2\pi} \angle \lim_{z \to \beta_j} \Psi_{z,j}(e^{it})^* \begin{pmatrix} I_p & -S(e^{it}) \\ -S(e^{it})^* & I_q \end{pmatrix} \Psi_{z,j}(e^{it}) \, dt \\
&\leq \angle \lim_{z \to \beta_j} \frac{1}{2\pi} \int_0^{2\pi} \Psi_{z,j}(e^{it})^* \begin{pmatrix} I_p & -S(e^{it}) \\ -S(e^{it})^* & I_q \end{pmatrix} \Psi_{z,j}(e^{it}) \, dt \\
&= \angle \lim_{z \to \beta_j} [\Psi_{z,j}, \Psi_{z,j}]_S \\
&= \angle \lim_{z \to \beta_j} \xi_j^* \frac{I_p - S(z)S(z)^*}{1 - |z|^2} \xi_j \leq \gamma_j = P_{jj}.
\end{aligned} \quad (8.16)$$

Since $P_S$ satisfies the Stein equation (8.4), its off diagonal blocks are equal to the corresponding blocks of $P$. Therefore, in view of (8.16), $P \geq P_S \geq 0$ and hence, by Lemma 1.3, $S \in \widehat{\mathcal{S}}(I_n, N, P, C)$. This completes the proof of the second assertion.

To verify the third assertion, assume that (8.7) holds and $P \geq 0$. Then, since $P$ is a solution of the Stein equation (8.4) by the first assertion, the $\widehat{\mathbf{aBIP}}(I_n, N, P, C)$ is defined. Let $S \in \widehat{\mathcal{S}}(I_n, N, P, C)$. Then, by Remark 1.1, the function $\widetilde{B}$ defined in (1.14) belongs to $\mathbf{H}_2^{n \times q}$. In view of the special choice (8.1) of $N$ and $C$,

$$\widetilde{B}(z) = \mathbf{col}\left(\frac{\eta_1^* - \xi_1^* S(z)}{z - \beta_1}, \ldots, \frac{\eta_m^* - \xi_m^* S(z)}{z - \beta_m}\right),$$

where the notation **col** signifies that the entries

$$\widetilde{B}_j(z) := \frac{\eta_j^* - \xi_j^* S(z)}{z - \beta_j} \in \mathbf{H}_2^{r_j \times q} \quad \text{for} \quad j = 1, \ldots, m$$

are stacked to form a block column matrix of size $(r_1 + \ldots + r_m) \times q = n \times q$. Thus, by Corollary 7.6 (for $j = 0$ and $k = 1$), which is applicable since $\widetilde{B}_j \in \mathbf{H}_2^{r_j \times q}$,

$$\angle \lim_{z \to \beta_j} \left(\eta_j^* - \xi_j^* S(z)\right) = \angle \lim_{z \to \beta_j} (z - \beta_j)\widetilde{B}_j(z) = 0 \quad (j = 1, \ldots, m),$$

which proves the first set of assertions in (8.13).

Since $S \in \widehat{\mathcal{S}}(I_n, N, P, C)$, the mvf $\mathbf{W}(z)$ defined by (3.12) (with $M = I_n$) is analytic in $\mathbb{D}$ and has positive semidefinite real part there (see Theorem 3.8). Since $\mathbf{W}$ takes the positive semidefinite value $\frac{1}{2}P$ at the origin, it admits a Riesz–Herglotz representation of the form

$$\mathbf{W}(z) = \frac{1}{2\pi}\int_0^{2\pi} \frac{e^{it} + z}{e^{it} - z}\, d\sigma(t), \tag{8.17}$$

where $d\sigma$ is a positive semidefinite $n \times n$–matrix–valued measure. By Lemma 7.1, the following limits exist

$$\angle \lim_{z \to \beta_j}(1 - z\bar{\beta}_j)\mathbf{W}(z) = \frac{\sigma(\{t_j\})}{\pi} \geq 0 \quad (\beta_j = e^{it_j}) \tag{8.18}$$

for $j = 1, \ldots, m$, where $\sigma(\{t_j\})$ denotes the matrix assigned to the point $\beta_j$ by the matrix–valued measure $\sigma$. Upon substituting the block decompositions (8.1), (8.2) and (8.3) of $C$, $G(z)$ and $H(z)$ into (3.12), we see that the diagonal blocks of $\mathbf{W}(z)$

$$\mathbf{W}_{jj}(z) = \frac{1}{2} \cdot \frac{\beta_j + z}{\beta_j - z}\gamma_j - \frac{z\beta_j}{(\beta_j - z)^2}\left(\xi_j^*\xi_j - \xi_j^*S(z)\eta_j\right).$$

Then, by (8.18),

$$\angle \lim_{z \to \beta_j}(1 - z\bar{\beta}_j)\mathbf{W}_{jj}(z) = \gamma_j - \angle \lim_{z \to \beta_j}\frac{\xi_j^*\xi_j - \xi_j^*S(z)\eta_j}{1 - z\bar{\beta}_j} = \frac{\sigma(\{t_j\})_{jj}}{\pi} \geq 0.$$

Thus,

$$\angle \lim_{z \to \beta_j}\frac{\xi_j^*\xi_j - \xi_j^*S(z)\eta_j}{1 - z\bar{\beta}_j} = \gamma_j - \frac{\sigma(\{t_j\})_{jj}}{\pi},$$

and, by Theorem 8.1,

$$\angle \lim_{z \to \beta_j}\xi_j^*\frac{I_p - S(z)S(z)^*}{\rho_z(z)}\xi_j = \angle \lim_{z \to \beta_j}\frac{\xi_j^*\xi_j - \xi_j^*S(z)\eta_j}{1 - z\bar{\beta}_j}$$

$$= \gamma_j - \frac{\sigma(\{t_j\})_{jj}}{\pi}. \tag{8.19}$$

This shows that $S$ meets the second set of conditions in (8.13) and hence completes the proof of the third assertion.

## 8. THE NEVANLINNA–PICK BOUNDARY PROBLEM

The first part of the final assertion is now selfevident from the second and third assertions. Moreover, since $N$ is a diagonal matrix, there exists a matrix $Q \in \mathbb{C}^{n \times r}$ with range $\mathcal{Q}$ such that the constraints (4.19) and (5.1) are in force and formulas (4.22) and (4.24) hold. □

We remark that in the third line of (8.16) we have equality:

LEMMA 8.3. *Let $S$ be a solution of the $\widehat{\mathbf{NPBP}}$. Then*

$$[P_S]_{jj} = \angle \lim_{z \to \beta_j} \xi_j^* \frac{I - S(z)S(z)^*}{\rho_z(z)} \xi_j. \tag{8.20}$$

PROOF. Let $S$ be a solution of the $\widehat{\mathbf{NPBP}}$. Then, by Theorem 8.2, $S \in \widehat{\mathcal{S}}(I_n, N, P, C)$ for $P$ defined by (8.14) and hence $S$ also belongs to $\widehat{\mathcal{S}}(I_n, N, P_S, C)$. Let $\mathbf{W}(z)$ be defined by (3.12) for this choice of $M$, $N$ and $P$ (i.e., with $M = I_n$ and $P = P_S$). Then the argument leading to the inequality (8.19) with $\gamma_j$ replaced by $[P_S]_{jj}$ yields the inequality

$$\angle \lim_{z \to \beta_j} \xi_j^* \frac{I_p - S(z)S(z)^*}{\rho_z(z)} \xi_j \leq [P_S]_{jj},$$

which together with (8.16) implies (8.20). □

Now we make a few remarks on the "equality case":

The **NPBP**: *Given $m$ distinct points $\beta_1, \ldots, \beta_m$ on $\mathbb{T}$ and given matrices $\xi_j \in \mathbb{C}^{p \times r_j}$ with $\text{rank} \, \xi_j = r_j$, $\eta_j \in \mathbb{C}^{q \times r_j}$ and $\gamma_j \in \mathbb{C}^{r_j \times r_j}$, find necessary and sufficient conditions which insure the existence of a Schur function $S \in \mathcal{S}^{p \times q}$ such that:*

$$\angle \lim_{z \to \beta_j} S(z)^* \xi_j = \eta_j \quad \text{and} \quad \angle \lim_{z \to \beta_j} \xi_j^* \frac{I - S(z)S(z)^*}{\rho_z(z)} \xi_j = \gamma_j \quad (j = 1, \ldots, m).$$

The next theorem is an analogue of Theorem 8.1: it shows that the **NPBP** is equivalent to the **aBIP**$(I_n, N, P, C)$.

THEOREM 8.4. *Let $N$, $C$ and $P$ be defined as in (8.1) and (8.14). Then:*

(1) *If $S$ is a solution of the **NPBP**, then $P$ is a positive semidefinite solution of the Stein equation (8.4) and $S$ is a solution of the **aBIP**$(I_n, N, P, C)$.*
(2) *If $P$ is a positive semidefinite solution of the Stein equation (8.4) and $S$ is a solution of the **aBIP**$(I_n, N, P, C)$, then $S$ also is a solution of the **NPBP**.*
(3) *The **NPBP** has a solution if and only if $P \geq 0$, condition (8.7) holds and, in addition,*

$$\text{Ker} \, P(I_n - \beta_j N) \bigcap \text{Ker} \, C \subseteq \text{Ker} \, P \quad \text{for} \quad j = 1, \ldots, m. \tag{8.21}$$

PROOF. Let $S$ be a solution of the **NPBP**. Then, by Theorem 8.1, $S \in \widehat{\mathcal{S}}(I_n, N, P, C)$. Moreover, by Lemma 8.3,

$$[P_S]_{jj} = \gamma_j = P_{jj}$$

and, since $P_S$ satisfies (8.4) by Remark 1.2,

$$[P_S]_{j\ell} = \frac{\xi_j^* \xi_\ell - \eta_j^* \eta_\ell}{1 - \beta_j \bar{\beta}_\ell}, \quad \text{for} \quad j \neq \ell.$$

Therefore, $P = P_S$ and by Lemma 1.3, $S$ is a solution of the **aBIP**$(I_n, N, P, C)$.

To verify the second assertion, note that the assumptions imply that $P_S = P$, and hence, by Theorem 8.1, that $S$ is a solution of the $\widehat{\text{NPBP}}$. Therefore, we can invoke Lemma 8.3 to conclude that in fact $S$ is a solution of the **NPBP**:

$$\angle \lim_{z \to \beta_j} \xi_j^* \frac{I_p - S(z)S(z)^*}{\rho_z(z)} \xi_j = [P_S]_{jj} = P_{jj} = \gamma_j.$$

Finally, the preceding analysis implies that the **NPBP** is solvable if and only if the **aBIP**$(I_n, N, P, C)$ is solvable. However, by Theorem 6.1, the **aBIP**$(I_n, N, P, C)$ has a solution if and only if

$$\operatorname{Ker} P(I_n - \beta_j N) \bigcap \operatorname{Ker} C = \operatorname{Ker} P \cap \operatorname{Ker} PN \cap \operatorname{Ker} C \quad \text{for} \quad j = 1, \ldots, m, \tag{8.22}$$

since $\beta_1, \ldots, \beta_m$ are the only points at which $G(\zeta)$ is not invertible. This serves to complete the proof, since (8.22) is equivalent to (8.21). □

Since the **NPBP** and the **aBIP**$(I_n, N, P, C)$ are equivalent, we may apply Theorem 6.3 to conclude that the **NPBP** has infinitely many solutions if $\nu < \min(p, q)$ (and so in particular, if the matrix $P$ defined in (8.14) is positive definite) and it has one solution if and only if

$$\nu := \operatorname{rank}(P + C_2^* C_2) - \operatorname{rank} P = \min(p, q). \tag{8.23}$$

An alternative condition to (8.21) for the solvability of the **NPBP** has been established in [**50**] for the case $p = q = 1$ and may be easily extended to the matrix–valued case as follows.

THEOREM 8.5. *Let $P$ be a positive semidefinite solution of the Stein equation (8.4) with coefficients specified by (8.1). Then:*

(1) *If $P$ is minimal in the sense that $P$ does not majorize any positive semidefinite block diagonal matrix $A$ of the form*

$$A = \operatorname{diag}\{A_1, \ldots, A_m\} \quad \text{with} \quad A_j \in \mathbb{C}^{r_j \times r_j}, \tag{8.24}$$

*then the **NPBP** is solvable and moreover, it is equivalent to the $\widehat{\text{NPBP}}$.*

(2) *If the **NPBP** is solvable and condition (8.23) holds, then $P$ is minimal in the sense described in the first part.*

PROOF. Let $P$ be a minimal positive semidefinite solution of (8.4) and let $S$ be any solution of the $\widehat{\text{NPBP}}$ (such a solution exists, by Theorem 8.1). By another application of Theorem 8.1, $S \in \widehat{\mathcal{S}}(I_n, N, P, C)$. Since $P_S$ satisfies the same Stein equation (8.4) as $P$, by Remark 1.2, $P - P_S$ is a positive semidefinite block diagonal matrix and

$$P - (P - P_S) = P_S \geq 0.$$

Therefore, by the presumed minimality, $P = P_S$ and hence, $S \in \mathcal{S}(I_n, N, P, C)$. By Theorem 8.4, $S$ is a solution of the **NPBP**, as claimed.

Suppose next that the **NPBP** is solvable and that (8.23) holds. Then the $\widehat{\text{NPBP}}$ has only one solution $S$ which is also the unique solution of the **NPBP** and therefore, $P_S = P$. Let

$$P' = P - A \geq 0,$$

where $A$ is a positive semidefinite block matrix of the form (8.24). Then $P'$ is a positive semidefinite solution of the Stein equation (8.4) and the $\widehat{\mathbf{aBIP}}(I_n, N, P', C)$ is defined and solvable. Since $P' \leq P$, it follows that

$$\widehat{S}(I_n, N, P', C) \subseteq \widehat{S}(I_n, N, P, C) = \{S\}.$$

Since the set on the left hand side is not empty, it contains only one element $S$ and therefore, $P_S \leq P'$. The last inequality together with relations $P' \leq P$ and $P = P_S$ implies $P' = P$. Therefore, $A = 0$, which is to say that $P$ is minimal. □

To complete the comparison with Theorem 1 in [**50**], we recall first that if $P$ is positive definite then, by Corollary 6.4, there are infinitely many solutions of the **NPBP**. Secondly, we give an independent proof of the fact that in the scalar case of the **NPBP** with distinct interpolation points $\beta_1, \ldots, \beta_m$ and singular $P \geq 0$ condition (8.23) is automatically met, i.e., the $\widehat{\mathbf{NPBP}}$ has a unique solution. More precisely, we establish the following result:

LEMMA 8.6. *Let $p = q = 1$, let $P$ be a positive semidefinite solution of the Stein equation (8.4) for the corresponding **NPBP** with distinct interpolation points $\beta_1, \ldots, \beta_m$ and $r_j = 1$ for $j = 1, \ldots, m$ in (8.1) and let $\nu$ be the integer defined in (8.23). Then:*

(1) *$P$ is invertible if and only if $\nu = 0$.*
(2) *$P$ is singular if and only if $\nu = 1$.*

PROOF. If $p = q = 1$, then $\operatorname{rank} C_2^* C_2 = 1$ and therefore, it is clear from (8.23) that $\nu$ is nonnegative and does not exceed one. Thus, $\nu = 1$ or $\nu = 0$ and consequently, in order to complete the proof, it suffices to establish the first assertion of the lemma.

In fact, since it is selfevident that $\operatorname{Ker} P = \{0\} \Rightarrow \nu = 0$, it remains only to prove the opposite implication. To this end, let us suppose that $\nu = 0$ and that $\dim(\operatorname{Ker} P) = k > 0$. Let $\ell = m - k$. Then, without loss of generality, we may assume that the interpolation points $\beta_1, \ldots, \beta_m$ have been ordered in such a way that the upper right hand $\ell \times \ell$ block of $P$ is invertible. Accordingly, let us introduce the block decompositions

$$P = \begin{pmatrix} P_{11} & P_{12} \\ P_{21} & P_{22} \end{pmatrix} \quad \text{and} \quad N = \begin{pmatrix} N_1 & 0 \\ 0 & N_2 \end{pmatrix},$$

where $P_{11}$ and $N_1$ are $\ell \times \ell$ matrices and let the columns of the block matrix $\begin{pmatrix} U \\ V \end{pmatrix}$ with components $U \in \mathbb{C}^{\ell \times k}$ and $V \in \mathbb{C}^{k \times k}$ span the kernel of $P$. Then, since $P_{11}$ is invertible and

$$P_{11} U + P_{12} V = 0_{\ell \times k},$$

it is readily seen that

$$\operatorname{Ker} V \subseteq \operatorname{Ker} \begin{pmatrix} U \\ V \end{pmatrix} = \{0\}.$$

Thus $V$ is invertible. Now, the assumption $\nu = 0$ together with (8.4) guarantees that

$$\operatorname{rank} P = \operatorname{rank}(P + C_2^* C_2) = \operatorname{rank}(N^* P N + C_1^* C_1),$$

i.e., that

$$\operatorname{Ker} P \bigcap \operatorname{Ker} C_2 = \operatorname{Ker} PN \bigcap \operatorname{Ker} C_1 = \operatorname{Ker} P. \qquad (8.25)$$

In particular, the subspace $\operatorname{Ker} P$ of $\mathbb{C}^m$ is $N$-invariant and therefore,

$$\begin{pmatrix} N_1 & 0 \\ 0 & N_2 \end{pmatrix} \begin{pmatrix} U \\ V \end{pmatrix} = \begin{pmatrix} U \\ V \end{pmatrix} R$$

for some $k \times k$ matrix $R$. But this in turn implies that

$$N_1 \left(UV^{-1}\right) = \left(UV^{-1}\right) N_2$$

and hence, upon letting $x_j$, $j = 1, \ldots, k$, denote the columns of $UV^{-1}$, that

$$\left(N_1 - \bar{\beta}_{\ell+j} I_\ell\right) x_j = 0 \quad \text{for} \quad j = 1, \ldots, k.$$

However, as the points $\beta_1, \ldots, \beta_m$ are distinct, the matrix $N_1 - \bar{\beta}_{\ell+j} I_\ell$ is invertible. Thus, $x_j = 0$ for $j = 1, \ldots, k$, i.e., $U = 0_{\ell \times k}$.

On the other hand, (8.25) implies that $\operatorname{Ker} C_2 \subseteq \operatorname{Ker} P$, i.e., that

$$C_2 \begin{pmatrix} U \\ V \end{pmatrix} = 0_{1 \times k},$$

which in turn reduces to

$$(\eta_{\ell+1}, \ldots, \eta_{\ell+k}) = 0_{1 \times k},$$

since $V$ is invertible. But this contradicts the fact that $|\eta_j| = 1$ for $j = 1, \ldots, n$. Thus, the choice $k > 0$ leads to a contradiction. Therefore, the only viable option is $k = 0$. $\square$

## 9. A multiple analogue of the Carathéodory–Julia theorem

In this section we begin the analysis of higher order nontangential derivatives. Our first objective is to extend Theorem 8.1. To this end, let $r$ be an integer satisfying $1 \leq r \leq \min(p, q)$, let $n = r(m+1)$ and let

$$M = I_n, \quad N = \begin{pmatrix} \bar{\beta} I_r & I_r & & \\ & \bar{\beta} I_r & \ddots & \\ & & \ddots & I_r \\ & & & \bar{\beta} I_r \end{pmatrix}, \quad \text{where } \beta \in \mathbb{T}, \qquad (9.1)$$

and

$$C = \begin{pmatrix} C_1 \\ C_2 \end{pmatrix} = \begin{pmatrix} \xi_0 & \cdots & \xi_m \\ \eta_0 & \cdots & \eta_m \end{pmatrix}, \quad \text{where } \xi_j \in \mathbb{C}^{p \times r}, \; \eta_j \in \mathbb{C}^{q \times r}. \qquad (9.2)$$

It is convenient to introduce the block shift matrix

$$T = \begin{pmatrix} 0 & I_r & & \\ & 0 & \ddots & \\ & & \ddots & I_r \\ & & & 0 \end{pmatrix} = (\delta_{i,j-1} I_r)_{i=0,\ldots,m}^{j=1,\ldots,m+1}, \qquad (9.3)$$

so that $N = \bar{\beta} I_n + T$. The mvf's $G(z)$ and $H(z)$ introduced via (1.4) and (1.12), respectively, now take the form

$$G(z) = (1 - z\bar{\beta}) I_n - zT \quad \text{and} \quad H(z) = (z - \beta) I_n - T^*.$$

Therefore,

$$G^{-1}(z) = \sum_{k=0}^{m} \frac{z^k T^k}{\rho_\beta(z)^{k+1}} = \begin{pmatrix} \frac{I_r}{\rho_\beta(z)} & \frac{zI_r}{\rho_\beta(z)^2} & \cdots & \frac{z^m I_r}{\rho_\beta(z)^{m+1}} \\ 0 & \frac{I_r}{\rho_\beta(z)} & \ddots & \vdots \\ \vdots & & \ddots & \frac{zI_r}{\rho_\beta(z)^2} \\ 0 & \cdots & 0 & \frac{I_r}{\rho_\beta(z)} \end{pmatrix} \quad (9.4)$$

and

$$H^{-1}(z) = \sum_{k=0}^{m} \frac{(T^*)^k}{(z-\beta)^{k+1}} = \begin{pmatrix} \frac{I_r}{z-\beta} & 0 & \cdots & 0 \\ \frac{I_r}{(z-\beta)^2} & \frac{I_r}{z-\beta} & \ddots & \vdots \\ \vdots & \ddots & \ddots & 0 \\ \frac{I_r}{(z-\beta)^{m+1}} & \cdots & \frac{I_r}{(z-\beta)^2} & \frac{I_r}{z-\beta} \end{pmatrix}. \quad (9.5)$$

Moreover, since

$$G(z) = -(z-\beta)\left(I_n + \frac{\beta T N^{-1}}{z-\beta}\right) N$$

and $N$ and $T$ commute, we may also write

$$G^{-1}(z) = \sum_{k=0}^{m} \frac{(-1)^{k+1}}{(z-\beta)^{k+1}} (\beta T)^k N^{-k-1}. \quad (9.6)$$

It is useful to introduce the matrix polynomials

$$\widetilde{H}(z) = (z-\beta)^{m+1} H^{-1}(z) = \sum_{k=0}^{m} (z-\beta)^k (T^*)^{m-k} \quad (9.7)$$

and

$$\widetilde{G}(z) = (z-\beta)^{m+1} G^{-1}(z) = \sum_{k=0}^{m} (-1)^{m-k+1} (z-\beta)^k (\beta T)^{m-k} N^{k-m-1} \quad (9.8)$$

and to note that

$$\widetilde{H}^{(j)}(\beta) = j!(T^*)^{m-j} \quad \text{and} \quad \widetilde{G}^{(j)}(\beta) = j!(-1)^{m-j+1} (\beta T)^{m-j} N^{j-m-1}. \quad (9.9)$$

We also introduce the kernel

$$\mathbf{L}_\omega(z) = \frac{1}{(m!)^2} \frac{\partial^{2m}}{\partial z^m \partial \bar{\omega}^m} \left( \widetilde{H}(z) C_1^* \Lambda_\omega(z) C_1 \widetilde{H}(\omega)^* \right), \quad (9.10)$$

where $\Lambda_\omega(z)$ is the reproducing kernel of $\mathcal{H}(S)$ defined in (2.5), and the mvf

$$\begin{aligned} \mathbf{V}(z) &= \frac{1}{m!} \frac{d^m}{dz^m} \widetilde{H}(z) C_1^* B(z) \\ &= \frac{1}{m!} \frac{d^m}{dz^m} \left( \widetilde{H}(z) C_1^* \left( C_1 - S(z) C_2 \right) G^{-1}(z) \right). \end{aligned} \quad (9.11)$$

In view of Proposition 2.3, $\mathbf{L}_\omega(z)$ is a positive kernel on $\mathbb{D} \times \mathbb{D}$. We shall partition it conformally with the block decomposition (9.7) of $\widetilde{H}(z)$ as

$$\mathbf{L}_\omega(z) = [\mathbf{L}_{ij}(z,\omega)]_{i,j=0}^m$$

with $r \times r$ matrix valued blocks $\mathbf{L}_{ij}(z,\omega)$. The lower right hand block $\mathbf{L}_{mm}(z,\omega)$ will play an important role in the subsequent analysis.

The next theorem is the main result of this section.

THEOREM 9.1. *Let* $S \in \mathcal{S}^{p \times q}$, *let* $C_1 \in \mathbb{C}^{p \times n}$, *let* $\widetilde{H}$, $\widetilde{G}$, $\mathbf{L}_\omega$ *and* $\mathbf{V}$ *be defined via* (9.7), (9.8), (9.10) *and* (9.11), *respectively, and let* $\mathbf{L}_{mm}(z, \omega)$ *be the lower right hand* $r \times r$ *block in* $\mathbf{L}_\omega(z)$. *The following are equivalent:*

(1) *For every point* $z$ *in a nontangential neighborhood* $\mathcal{U}_\beta$ *of* $\beta$,

$$\|\mathbf{L}_z(z)\| < k < \infty. \tag{9.12}$$

(2) *The nontangential limit*

$$P_\mathbf{L} := \angle \lim_{z, \omega \to \beta} \mathbf{L}_\omega(z) \tag{9.13}$$

*exists.*

(3) *For every point* $z$ *in a nontangential neighborhood* $\mathcal{U}_\beta$ *of* $\beta$,

$$\|\mathbf{L}_{mm}(z, z)\| < k_1 < \infty. \tag{9.14}$$

(4) *The nontangential limit* $\angle \lim_{z, \omega \to \beta} \mathbf{L}_{mm}(z, \omega)$ *exists.*

*Moreover, if any one (and hence every one) of the preceding conditions is in force, then*

(a) $\angle \lim_{z \to \beta} \dfrac{1}{m!} \dfrac{d^m}{dz^m} \left( \widetilde{H}(z) C_1^* S(z) \right) = C_2^*.$ \hfill (9.15)

(b) $\angle \lim_{z \to \beta} \dfrac{1}{m!} \dfrac{d^m}{dz^m} \left( S(z) C_2 \widetilde{G}(z) \right) = -C_1 N^{-1}.$ \hfill (9.16)

(c) $\displaystyle\sum_{j=0}^{m} (T^*)^{m-j} \left( C_1^* C_1 - C_2^* C_2 \right) (-\beta T)^j N^{-j-1} = 0.$ \hfill (9.17)

(d) *The nontangential limit*

$$P_\mathbf{V} := \angle \lim_{z \to \beta} \mathbf{V}(z) \tag{9.18}$$

*exists and moreover,*

$$P_\mathbf{V} = P_\mathbf{L} = P_S, \tag{9.19}$$

*where* $P_S$ *is the positive semidefinite matrix associated with* $S$ *via* (1.10) *(and hence* $P_\mathbf{V}$ *is positive semidefinite).*

(e) *The columns of the function* $B(\zeta) = (C_1 - S(\zeta) C_2) G^{-1}(\zeta)$ *belong to* $\mathcal{H}(S)$ *and*

$$\langle By, Bx \rangle_{\mathcal{H}(S)} = x^* P_S y$$

*for every choice of* $x, y \in \mathbb{C}^n$. *In particular,* $B \in \mathbf{H}_2^{p \times n}$.

(f) *The function* $\widetilde{B}(\zeta) = H(\zeta)^{-1} \left( C_2^* - C_1^* S(\zeta) \right)$ *belongs to* $\mathbf{H}_2^{n \times q}$.

We remark that (9.15) serves to define the matrix $C_2$. Then $C_1$ and $C_2$ are subject to (9.17) and the nontangential limit in (9.16) exists and is equal to $-C_1 N^{-1}$. Moreover, the Lyapunov-Stein equation (1.7) admits a solution if and only if the condition (9.17) is in force. Under the stronger condition (9.12), it admits a positive semidefinite solution. We shall discuss solutions of (1.7) in more detail in Sections

10 and 11. An example in which (9.12) fails and the limit $P_{\mathbf{V}}$ exists, but is not Hermitian, is furnished just after the statement of Theorem 9.8.

Furthermore, for $m = 0$ and for the choice $C_1 = \xi$ and $C_2 = \eta$, the equalities (9.15), (9.16) and (9.17) coincide with (8.10) and the two relations in (8.11), respectively, whereas the nontangential limits (9.18) and (9.13) coincide with the nontangential limits in (8.12) and (8.9). Thus, Theorem 8.1 is a special case of Theorem 9.1 which in turn is an elaborate generalization of Lemma 8.3 of [**23**]. Formula (9.59) in the former plays the role of formula (8.6) in the latter. A quick look at the relative complexities of these two formulas is a good indicator of the extra effort required to establish this generalization. The proof of Theorem 9.1 will be given below after a number of preliminary lemmas.

LEMMA 9.2. *Let $G^{-1}$ and $\widetilde{H}$ be defined by (9.4) and (9.7), respectively. Then*

$$\frac{1}{m!}\frac{\partial^m}{\partial \bar{\omega}^m}\left(\frac{\widetilde{H}(\omega)^*}{\rho_\omega(z)}\right) = G(z)^{-1}\left(\frac{\rho_\beta(z)}{\rho_\omega(z)}\right)^{m+1}. \qquad (9.20)$$

PROOF. We begin with the resolvent like identities

$$H(\omega)^{-*} - zG(z)^{-1} = \rho_\omega(z)G(z)^{-1}MH(\omega)^{-*} = \rho_\omega(z)G(z)^{-1}H(\omega)^{-*},$$

which follow readily from the definitions (1.4) and (1.12). Therefore, by (9.7),

$$\frac{\widetilde{H}(\omega)^*}{\rho_\omega(z)} = G(z)^{-1}\left(\widetilde{H}(\omega)^* + \frac{z(\bar{\omega}-\bar{\beta})^{m+1}}{\rho_\omega(z)}I_n\right). \qquad (9.21)$$

By Leibnitz's rule,

$$\frac{1}{m!}\frac{\partial^m}{\partial \bar{\omega}^m}\left(\frac{(\bar{\omega}-\bar{\beta})^{m+1}}{\rho_\omega(z)}\right) = \sum_{j=0}^{m}\binom{m+1}{j+1}\frac{z^j(\bar{\omega}-\bar{\beta})^{j+1}}{\rho_\omega(z)^{j+1}}$$

and, since $\dfrac{d^m}{d\bar{\omega}^m}\widetilde{H}(\omega)^* = m!I_n$, it follows from (9.21) that

$$\begin{aligned}\frac{1}{m!}\frac{\partial^m}{\partial \bar{\omega}^m}\left(\frac{\widetilde{H}(\omega)^*}{\rho_\omega(z)}\right) &= G(z)^{-1}\left(1 + z\sum_{j=0}^{m}\binom{m+1}{j+1}\frac{z^j(\bar{\omega}-\bar{\beta})^{j+1}}{\rho_\omega(z)^{j+1}}\right) \\ &= G(z)^{-1}\sum_{j=0}^{m+1}\binom{m+1}{j}\left(\frac{z(\bar{\omega}-\bar{\beta})}{\rho_\omega(z)}\right)^j \\ &= G(z)^{-1}\left(1 + \frac{z(\bar{\omega}-\bar{\beta})}{\rho_\omega(z)}\right)^{m+1} = G(z)^{-1}\left(\frac{\rho_\beta(z)}{\rho_\omega(z)}\right)^{m+1}.\end{aligned}$$

□

LEMMA 9.3. *Let* $S \in \mathcal{S}^{p \times q}$, $C_1 \in \mathbb{C}^{p \times n}$ *and let* $\widetilde{H}$ *be defined by* (9.7). *Then the following statements are equivalent:*

(1) $S(z)$ *satisfies* (9.15).

(2) $\dfrac{1}{j!} \angle \lim_{z \to \beta} \dfrac{d^j}{dz^j} \left( \widetilde{H}(z) C_1^* S(z) \right) = (T^*)^{m-j} C_2^*$ (*for* $j = 0, \ldots, m$). (9.22)

(3) $\dfrac{1}{m!} \angle \lim_{z \to \beta} \dfrac{\partial^m}{\partial z^m} \left( \dfrac{\widetilde{H}(z) C_1^* S(z)}{\rho_\zeta(z)} \right) = G(\zeta)^{-*} C_2^*$ ($\forall\, \zeta \in \mathbb{C} \setminus \{\beta\}$). (9.23)

(4) $\angle \lim_{z \to \beta} (z - \beta) H(z)^{-1} \left( C_1^* S(z) - C_2^* \right) = 0.$ (9.24)

PROOF. **(1)** $\Rightarrow$ **(2)**. Let (9.15) be in force. Then, by Leibnitz's rule,

$$\frac{1}{m!} \angle \lim_{z \to \beta} \frac{d^m}{dz^m} \left( (z - \beta) \widetilde{H}(z) C_1^* S(z) \right) = \frac{1}{(m-1)!} \angle \lim_{z \to \beta} \frac{d^{m-1}}{dz^{m-1}} \left( \widetilde{H}(z) C_1^* S(z) \right). \quad (9.25)$$

On the other hand, making use of (9.15) and the equality

$$(z - \beta) \widetilde{H}(z) = T^* \widetilde{H}(z) + (z - \beta)^{m+1} I_n, \quad (9.26)$$

which follows readily from (9.7), we obtain

$$\frac{1}{m!} \angle \lim_{z \to \beta} \frac{d^m}{dz^m} \left( (z - \beta) \widetilde{H}(z) C_1^* S(z) \right)$$
$$= T^* C_2^* + \frac{1}{m!} \angle \lim_{z \to \beta} \frac{d^m}{dz^m} \left( (z - \beta)^{m+1} C_1^* S(z) \right). \quad (9.27)$$

Since $C_1^* S(z)$ belongs to $\mathbf{H}_2^{n \times q}$, the second term on the right hand side of (9.27) is equal to zero by Corollary 7.6. Then, a comparison of (9.25) and (9.27) leads to the formula

$$\frac{1}{(m-1)!} \angle \lim_{z \to \beta} \frac{d^{m-1}}{dz^{m-1}} \left( \widetilde{H}(z) C_1^* S(z) \right) = T^* C_2^*,$$

which proves (9.22) for $j = m - 1$. Using similar arguments we obtain (9.22) recursively for all integers $j = m - 2, \ldots, 1, 0$.

**(2)** $\Rightarrow$ **(3)**. Making use of (9.22), (9.4) and Leibnitz's rule, we obtain

$$\frac{1}{m!} \angle \lim_{z \to \beta} \frac{\partial^m}{\partial z^m} \left( \frac{\widetilde{H}(z) C_1^* S(z)}{\rho_\zeta(z)} \right)$$
$$= \angle \lim_{z \to \beta} \sum_{j=0}^{m} \frac{1}{(m-j)!} \left( \widetilde{H}(z) C_1^* S(z) \right)^{(m-j)} \frac{1}{j!} \left( \frac{1}{\rho_\zeta(z)} \right)^{(j)}$$
$$= \sum_{j=0}^{m} (T^*)^j C_2^* \frac{\bar{\zeta}^j}{\rho_\beta(z)^{j+1}}$$
$$= \left( \sum_{j=0}^{m} \frac{\zeta^j T^j}{\rho_\beta(z)^{j+1}} \right)^* C_2^* = G(\zeta)^{-*} C_2^*.$$

**(3)** $\Rightarrow$ **(1)**. Setting $\zeta = 0$ in (9.23), we get (9.15).

## 9. A MULTIPLE ANALOGUE OF THE CARATHÉODORY–JULIA THEOREM

$(\mathbf{2}) \Leftrightarrow (\mathbf{4})$. Set

$$F(z) = \widetilde{H}(z)C_1^* S(z) \quad \text{and} \quad F_j = (T^*)^{m-j} C_2^* \quad (j=0,\ldots,m). \tag{9.28}$$

Then, by (9.7),

$$\widetilde{H}(z)C_2^* = F_0 + (z-\beta)F_1 + \cdots + (z-\beta)^m F_m$$

and hence we can rewrite (9.24) and (9.22) as (7.21) and (7.22), respectively. But the two latter relations are equivalent by Corollary 7.9, since a mvf $F$ of the form (9.28) is analytic in $\mathbb{D}$. $\square$

The next lemma can be proved in much the same way.

LEMMA 9.4. *Let $S \in \mathcal{S}^{p \times q}$, $C_2 \in \mathbb{C}^{q \times n}$ and let $\widetilde{G}$ be defined by (9.8). Then the following statements are equivalent:*

(1) $S(z)$ *satisfies* (9.16).

(2) $\displaystyle \angle \lim_{z \to \beta} \frac{1}{j!} \frac{d^j}{dz^j}\left(S(z)C_2\widetilde{G}(z)\right) = (-1)^{m-j+1} C_1 (\beta T)^{m-j} N^{j-m-1}$ \hfill (9.29)

(*for* $j = 0, \ldots, m$).

(3) $\displaystyle \angle \lim_{z \to \beta} \frac{1}{m!} \frac{\partial^m}{\partial z^m}\left(\frac{S(z)C_2\widetilde{G}(z)}{\rho_\zeta(z)}\right) = C_1 H(\zeta)^{-*} \quad (\forall\, \zeta \in \mathbb{C}\setminus\{\beta\}).$

(4) $\displaystyle \angle \lim_{z \to \beta} (z-\beta)\left(S(z)C_2 - C_1\right)G(z)^{-1} = 0.$ \hfill (9.30)

Let $S \in \mathcal{S}^{p \times q}$, $C_1 \in \mathbb{C}^{p \times n}$ and let $\widetilde{H}$ be given by (9.7). The mvf

$$\boldsymbol{\Psi}_z(\zeta) = \begin{pmatrix} \boldsymbol{\Psi}_z^1(\zeta) \\ \boldsymbol{\Psi}_z^2(\zeta) \end{pmatrix} = \frac{1}{m!} \frac{\partial^m}{\partial \bar{z}^m}\left(\begin{pmatrix} I_p \\ S(z)^* \end{pmatrix} \frac{C_1 \widetilde{H}(z)^*}{\rho_z(\zeta)}\right) \tag{9.31}$$

which is defined and belongs to $\mathbf{H}_2^{(p+q) \times n}$ as a (rational) function of $\zeta$ for each fixed choice of the parameter $z \in \mathbb{D}$, will play an important role in the subsequent analysis. It is clear from (9.31) that

$$\boldsymbol{\Psi}_z^1(\zeta) = \frac{1}{m!} \frac{\partial^m}{\partial \bar{z}^m}\left(\frac{C_1 \widetilde{H}(z)^*}{\rho_z(\zeta)}\right) \quad \text{and} \quad \boldsymbol{\Psi}_z^2(\zeta) = \frac{1}{m!} \frac{\partial^m}{\partial \bar{z}^m}\left(\frac{S(z)^* C_1 \widetilde{H}(z)^*}{\rho_z(\zeta)}\right). \tag{9.32}$$

LEMMA 9.5. *Let $S \in \mathcal{S}^{p \times q}$ and let $\boldsymbol{\Psi}_z$ be defined by (9.31). Then:*

(1) $\boldsymbol{\Psi}_z^2 = \underline{p}S^*\boldsymbol{\Psi}_z^1$, *where $\underline{p}$ denotes the orthogonal projection of $L_2^p(\mathbb{T})$ onto $\mathbf{H}_2^p$.*

(2) *For every point $\zeta \in \mathbb{C}\setminus\{\beta\}$,*

$$\angle \lim_{z \to \beta} \boldsymbol{\Psi}_z^1(\zeta) = C_1 G(\zeta)^{-1}. \tag{9.33}$$

(3) *If, moreover, the nontangential limit (9.15) exists, then*

$$\angle \lim_{z \to \beta} \boldsymbol{\Psi}_z^2(\zeta) = C_2 G(\zeta)^{-1} \quad (\zeta \in \mathbb{C}\setminus\{\beta\}). \tag{9.34}$$

PROOF. Since
$$\underline{p}\left(S^*\frac{\partial^j}{\partial \bar{z}^j}\left(\frac{1}{\rho_z}\right)\right) = \frac{\partial^j}{\partial \bar{z}^j}\left(\frac{S(z)^*}{\rho_z}\right),$$
(see e.g., [**23**, Lemma 6.1]), it follows from (9.31) that

$$\begin{aligned}
\underline{p}S^*\Psi_z^1 &= \frac{1}{m!}\underline{p}S^*\frac{\partial^m}{\partial \bar{z}^m}\left(\frac{C_1\widetilde{H}(z)^*}{\rho_z}\right) \\
&= \frac{1}{m!}\underline{p}S^*\sum_{j=0}^{m}\binom{m}{j}\left(\frac{\partial^j}{\partial \bar{z}^j}\left(\frac{1}{\rho_z}\right)\right)C_1\widetilde{H}^{(m-j)}(z)^* \\
&= \frac{1}{m!}\sum_{j=0}^{m}\binom{m}{j}\underline{p}\left(S^*\frac{\partial^j}{\partial \bar{z}^j}\left(\frac{1}{\rho_z}\right)\right)C_1\widetilde{H}^{(m-j)}(z)^* \\
&= \frac{1}{m!}\sum_{j=0}^{m}\binom{m}{j}\left(\frac{\partial^j}{\partial \bar{z}^j}\left(\frac{S(z)^*}{\rho_z}\right)\right)C_1\widetilde{H}^{(m-j)}(z)^* \\
&= \frac{1}{m!}\frac{\partial^m}{\partial \bar{z}^m}\left(\frac{S(z)^*C_1\widetilde{H}(z)^*}{\rho_z}\right) = \Psi_z^2,
\end{aligned}\qquad(9.35)$$

which proves the first assertion of lemma. By (9.20) and (9.32),
$$\Psi_z^1(\zeta) = G(\zeta)^{-1}\left(\frac{\rho_\beta(\zeta)}{\rho_z(\zeta)}\right)^{m+1}$$
and upon taking limits in the last equality as $z \to \beta$ we obtain (9.33). The last statement of the lemma follows readily from (9.23).

LEMMA 9.6. *Let* $S \in \mathcal{S}^{p\times q}$, $x, y \in \mathbb{C}^n$, $z, \omega \in \mathbb{D}$ *and let*
$$B_z(\zeta) = (I_p,\ -S(\zeta))\,\Psi_z(\zeta) = \frac{1}{m!}\frac{\partial^m}{\partial \bar{z}^m}\left(\Lambda_z(\zeta)C_1\widetilde{H}(z)^*\right). \qquad(9.36)$$
*Then the function* $B_z(\zeta)x$ *belongs to the space* $\mathcal{H}(S)$ *and*
$$\langle B_\omega y,\ B_z x\rangle_{\mathcal{H}(S)} = \langle B_\omega y,\ \Psi_z^1 x\rangle_{\mathbf{H}_2^p} = [\Psi_\omega y,\ \Psi_z x]_S = x^*\mathbf{L}_\omega(z)y. \qquad(9.37)$$
*Moreover, if the nontangential limit* (9.15) *exists, then*
$$\angle\lim_{z\to\beta} B_z(\zeta) = (I_p,\ -S(\zeta))\,CG(\zeta)^{-1} \qquad (\forall\ \zeta \in \mathbb{C}\backslash\{\beta\}). \qquad(9.38)$$

PROOF. By (9.36) and statement (1) in Lemma 9.5,
$$B_z = (I_p,\ -S)\,\Psi_z = (I - S\underline{p}S^*)\Psi_z^1. \qquad(9.39)$$
It is known (see e.g., [**25**, Lemma 6.2]) that for every choice of $h \in \mathbf{H}_2^p$, the function $(I - S\underline{p}S^*)h$ belongs to the space $\mathcal{H}(S)$ and
$$\langle g,\ (I - S\underline{p}S^*)h\rangle_{\mathcal{H}(S)} = \langle g,\ h\rangle_{\mathbf{H}_2^p} \qquad (\forall\ g \in \mathcal{H}(S)). \qquad(9.40)$$
Therefore, since the function $\Psi_z^1 x$ belongs to $\mathbf{H}_2^p$ for every choice of the vector $x \in \mathbb{C}^n$, $B_z x \in \mathcal{H}(S)$ by (9.39), whereas (9.40) leads to
$$\langle B_\omega y,\ B_z x\rangle_{\mathcal{H}(S)} = \langle B_\omega y,\ (I - S\underline{p}S^*)\Psi_z^1 x\rangle_{\mathcal{H}(S)} = \langle B_\omega y,\ \Psi_z^1 x\rangle_{\mathbf{H}_2^p}.$$

Next, in view of (9.35),

$$(-S^*, \ I_q) \, \boldsymbol{\Psi}_\omega y = (-S^*, \ I_q) \begin{pmatrix} I \\ \underline{pS^*} \end{pmatrix} \Psi_z^1 y = -(I - \underline{p}) S^* \Psi_z^1 y \in (\mathbf{H}_2^q)^\perp$$

and thus, as $\Psi_z^2 x \in \mathbf{H}_2^q$,

$$\langle (-S^*, \ I_q) \, \boldsymbol{\Psi}_\omega y, \ \Psi_z^2 x \rangle_{\mathbf{H}_2^q} = 0.$$

Therefore,

$$\begin{aligned} \left[ \boldsymbol{\Psi}_\omega y, \ \boldsymbol{\Psi}_z x \right]_S &= \left\langle \begin{pmatrix} I_p & -S \\ -S^* & I_q \end{pmatrix} \boldsymbol{\Psi}_\omega y, \ \boldsymbol{\Psi}_z x \right\rangle_{L_2^{p+q}} \\ &= \langle (I_p, -S) \boldsymbol{\Psi}_\omega y, \ \Psi_z^1 x \rangle_{L_2^p} + \langle (-S^*, I_q) \boldsymbol{\Psi}_\omega y, \ \Psi_z^2 x \rangle_{L_2^q} \\ &= \langle B_\omega y, \ \Psi_z^1 x \rangle_{L_2^p}, \end{aligned}$$

which completes the proof of (9.37) except for the last equality. But that follows easily from the reproducing kernel formula

$$\langle f, \ B_z x \rangle_{\mathcal{H}(S)} = \frac{1}{m!} \frac{d^m}{dz^m} \left( x^* \widetilde{H}(z) C_1^* f(z) \right), \qquad (9.41)$$

which is valid for every choice of $f \in \mathcal{H}(S)$, by choosing $f = B_\omega y$. Finally, (9.38) follows from (9.33) and (9.34), which, in turn, hold by Lemma 9.5. □

LEMMA 9.7. *Conditions* (9.12) *and* (9.14) *are equivalent.*

PROOF. Since $\|\mathbf{L}_{mm}(z,z)\| \leq \|\mathbf{L}_z(z)\|$ for every point $z$ at which $\mathbf{L}_z(z)$ is defined, it remains to show that (9.14) implies (9.12). To this end, introduce the matrix

$$E = (I_r, \ 0, \ \ldots, \ 0) \in \mathbb{C}^{r \times n} \qquad (9.42)$$

and note that

$$ET^\ell (T^*)^k = \begin{cases} ET^{\ell-k} & \text{if } \ell \geq k, \\ 0 & \text{if } \ell < k, \end{cases} \quad (0 \leq \ell, k \leq m). \qquad (9.43)$$

Making use of (9.43) and (9.7), we can reexpress $ET^\ell \widetilde{H}(z) C_1^*$ in terms of the $\mathbb{C}^{r \times p}$-valued polynomials

$$\mathcal{A}_\ell(z) = \sum_{k=0}^{\ell} (z - \beta)^k ET^k C_1^* \quad (\ell = 0, \ldots, m) \qquad (9.44)$$

as

$$\begin{aligned} ET^\ell \widetilde{H}(z) C_1^* &= ET^\ell \sum_{k=0}^{m} (z - \beta)^k (T^*)^{m-k} C_1^* \\ &= (z - \beta)^{m-\ell} \sum_{k=0}^{\ell} (z - \beta)^k ET^k C_1^* \\ &= (z - \beta)^{m-\ell} \mathcal{A}_\ell(z) \quad (\ell = 0, \ldots, m). \end{aligned}$$

The latter equalities mean that the $\ell$-th block row of the matrix $\widetilde{H}(z)C_1^*$ is equal to $(z-\beta)^{m-\ell}\mathcal{A}_\ell(z)$, i.e., that

$$\widetilde{H}(z)C_1^* = \begin{pmatrix} (z-\beta)^m \mathcal{A}_0(z) \\ \vdots \\ (z-\beta)\mathcal{A}_{m-1}(z) \\ \mathcal{A}_m(z) \end{pmatrix}. \tag{9.45}$$

Upon substituting the latter representation into (9.10) we conclude that the block entries $\mathbf{L}_{ij}(z,w)$ of $\mathbf{L}_\omega(z)$ can be represented as

$$\mathbf{L}_{ij}(z,w) = \frac{1}{(m!)^2} \frac{\partial^{2m}}{\partial z^m \partial \bar{\omega}^m} \left[ (z-\beta)^{m-i} \mathcal{A}_i(z) \Lambda_\omega(z) \mathcal{A}_j(\omega)^* (\bar{\omega}-\bar{\beta})^{m-j} \right]. \tag{9.46}$$

The next step is to invoke the representation (9.46) to show that

$$\text{if } \sup_{z,w \in \mathcal{U}_\beta} \|\mathbf{L}_{\ell,\ell}(z,\omega)\| < \infty, \text{ then } \sup_{z,w \in \mathcal{U}_\beta} \|\mathbf{L}_{\ell-1,\ell-1}(z,w)\| < \infty \tag{9.47}$$

in each nontangential neighborhood $\mathcal{U}_\beta$ of $\beta$. Indeed, the inequality

$$\mathcal{A}_{\ell-1}(z)\Lambda_\omega(z)\mathcal{A}_{\ell-1}(\omega)^* \preceq 2\mathcal{A}_\ell(z)\Lambda_\omega(z)\mathcal{A}_\ell(\omega)^*$$
$$+ 2(z-\beta)^\ell (\bar{\omega}-\bar{\beta})^\ell ET^\ell C_1^* \Lambda_\omega(z) C_1 (T^*)^\ell E^*$$

holds for every $\ell = 1, \ldots, m$ and follows easily from the decomposition

$$\mathcal{A}_{\ell-1}(z) = \mathcal{A}_\ell(z) - (z-\beta)^\ell ET^\ell C_1^*$$

and Proposition 2.1. Therefore, by Proposition 2.3 and (9.46),

$$\mathbf{L}_{\ell-1,\ell-1}(z,\omega) \preceq 2\mathbf{K}_\ell^1(z,\omega) + 2\mathbf{K}_\ell^2(z,\omega), \tag{9.48}$$

where

$$\mathbf{K}_\ell^1(z,\omega) = \frac{1}{(m!)^2} \frac{\partial^{2m}}{\partial z^m \partial \bar{\omega}^m} \left( (z-\beta)^{m-\ell+1} (\bar{\omega}-\bar{\beta})^{m-\ell+1} \mathcal{A}_\ell(z) \Lambda_\omega(z) \mathcal{A}_\ell(\omega)^* \right)$$

and

$$\mathbf{K}_\ell^2(z,\omega) = \frac{1}{(m!)^2} \frac{\partial^{2m}}{\partial z^m \partial \bar{\omega}^m} \left( (z-\beta)^{m+1} (\bar{\omega}-\bar{\beta})^{m+1} ET^\ell C_1^* \Lambda_\omega(z) C_1 (T^*)^\ell E^* \right).$$

If $\mathbf{L}_{\ell,\ell}(z,\omega)$ is uniformly bounded on $\mathcal{U}_\beta \times \mathcal{U}_\beta$, then $\mathbf{K}_\ell^1$ is also uniformly bounded for $z,\omega \in \mathcal{U}_\beta$, by Corollary 7.11. Moreover, by Lemma 7.7, applied first with $F(z) = I_p$ and then with $F(z) = S(z)$, we conclude that $\mathbf{K}_\ell^2(z,\omega)$ tends to zero, as $z$ and $\omega$ tend nontangentially to $\beta$. Therefore, since each nontangential neighborhood $\mathcal{U}_\beta$ of $\beta$ sits inside a closed cone in $\mathbb{D}$, $\mathbf{K}_\ell^2(z,\omega)$ is uniformly bounded for $z,\omega \in \mathcal{U}_\beta$. Now (9.47) follows from the inequality (9.48), by Proposition 2.2.

It follows from (9.14) by Cauchy's inequality, that $\mathbf{L}_{m,m}(z,\omega)$ is uniformly bounded on $\mathcal{U}_\beta \times \mathcal{U}_\beta$. Then by (9.47) all the diagonal blocks $\mathbf{L}_{\ell,\ell}(z,w)$ of the kernel $\mathbf{L}_\omega(z)$ are uniformly bounded on $\mathcal{U}_\beta \times \mathcal{U}_\beta$. Thus, as the kernel $\mathbf{L}(z,\omega)$ is positive on $\mathbb{D} \times \mathbb{D}$, it follows that its off diagonal blocks are also uniformly bounded. Therefore, $\mathbf{L}(z,\omega)$ is uniformly bounded for every pair of points $z$ and $\omega$ from $\mathcal{U}_\beta$, which clearly implies (9.12). $\square$

## 9. A MULTIPLE ANALOGUE OF THE CARATHÉODORY–JULIA THEOREM

**Proof of Theorem 9.1:** The equivalence (1) $\Leftrightarrow$ (3) has been established in Lemma 9.7. The implications (2) $\Rightarrow$ (1), (4) $\Rightarrow$ (3) and (2) $\Rightarrow$ (4) are selfevident. Thus, to complete the proof of the theorem, it remains to show that if the constraint (9.12) is in force, then the nontangential limit (9.13) exists and that the conclusions (a)–(f) all hold.

In view of (9.12) and (9.37),

$$\|B_z x\|^2_{\mathcal{H}(S)} = x^* \mathbf{L}_z(z) x < k \|x\|^2 \qquad (\forall\, x \in \mathbb{C}^n,\ z \in \mathcal{U}_\beta). \tag{9.49}$$

Let $R_0 : \mathbf{H}_2^{p \times q} \to \mathbf{H}_2^{p \times q}$ be the backward shift operator defined (in accordance with (4.2)) by the rule

$$(R_0 F)(z) = \frac{F(z) - F(0)}{z}.$$

Then, since $R_0 S y = \frac{S(z) - S(0)}{z} y$ belongs to $\mathcal{H}(S)$ and $\|R_0 S y\|_{\mathcal{H}(S)} \leq \|y\|$ for every choice of $y \in \mathbb{C}^q$ (for a proof see e.g., [23, Theorem 2.3]), we may choose $f = R_0 S y$ in (9.41) and invoke Cauchy's inequality to obtain the estimate

$$\left| \frac{1}{m!} \frac{d^m}{dz^m} \left( x^* \widetilde{H}(z) C_1^* (R_0 S)(z) y \right) \right| = |\langle R_0 S y, B_z x \rangle_{\mathcal{H}(S)}|$$
$$\leq \|R_0 S y\|_{\mathcal{H}(S)} \|B_z x\|_{\mathcal{H}(S)} < k^{\frac{1}{2}} \|x\| \|y\|.$$

Since $x$ and $y$ are arbitrary, this implies that

$$\left\| \frac{d^m}{dz^m} \left( \widetilde{H}(z) C_1^* (R_0 S)(z) \right) \right\| < k_1 \qquad (z \in \mathcal{U}_\beta).$$

By Remark 7.10,

$$\left\| \frac{d^{m-1}}{dz^{m-1}} \left( \widetilde{H}(z) C_1^* (R_0 S)(z) \right) \right\| < k_2 \qquad (z \in \mathcal{U}_\beta)$$

for some finite constant $k_2 \geq k_1$ and hence, by Leibnitz's rule, we conclude from the last two bounds that

$$\left\| \frac{d^m}{dz^m} \left( \widetilde{H}(z) C_1^* (S(z) - S(0)) \right) \right\| \leq \left\| z \frac{d^m}{dz^m} \left( \widetilde{H}(z) C_1^* (R_0 S)(z) \right) \right\|$$
$$+ m \left\| \frac{d^{m-1}}{dz^{m-1}} \left( \widetilde{H}(z) C_1^* (R_0 S)(z) \right) \right\|$$
$$< (m+1) k_2. \tag{9.50}$$

Since

$$\frac{d^m}{dz^m} \left( \widetilde{H}(z) C_1^* S(0) \right) = m! C_1^* S(0),$$

it follows from (9.50) that

$$\left\| \frac{d^m}{dz^m} \left( \widetilde{H}(z) C_1^* S(z) \right) \right\| < k_3 \qquad (z \in \mathcal{U}_\beta) \tag{9.51}$$

for some finite constant $k_3$. In view of assumption (9.12) and the derived bound (9.51), there exists a sequence of points $\alpha_i \in \mathcal{U}_\beta$ tending to $\beta$ such that the limits

$$\lim_{\alpha_i \to \beta} \mathbf{L}_{\alpha_i}(\alpha_i) = P_{\mathbf{L}} \in \mathbb{C}^{n \times n} \tag{9.52}$$

and

$$\lim_{\alpha_i \to \beta} \frac{1}{m!} \frac{d^m}{dz^m} \left( \widetilde{H}(z) C_1^* S(z) \right)_{z = \alpha_i} = C_2^* \in \mathbb{C}^{n \times p}$$

exist. By (the proof of **(2)** ⇒ **(3)** in) Lemma 9.3, the existence of the last limit implies the existence of the limit

$$\frac{1}{m!} \lim_{\alpha_i \to \beta} \frac{\partial^m}{\partial z^m} \left( \frac{\widetilde{H}(z) C_1^* S(z)}{1 - z\bar{\zeta}} \right)_{z=\alpha_i} = G(\zeta)^{-*} C_2^*$$

for every point $\zeta \in \mathbb{C} \setminus \{\beta\}$. It is convenient to break up the rest of the proof into steps.

**Step 1.** *The integral* (1.11) *converges to a matrix* $P_S$ *which satisfies*

$$P_S := \left[ \begin{pmatrix} C_1 \\ C_2 \end{pmatrix} G(\zeta)^{-1}, \begin{pmatrix} C_1 \\ C_2 \end{pmatrix} G(\zeta)^{-1} \right]_S \leq P_{\mathbf{L}}, \qquad (9.53)$$

*where* $P_{\mathbf{L}}$ *is the matrix defined via* (9.52).

**Proof of Step 1:** By (9.37),

$$\mathbf{L}_\omega(z) = [\mathbf{\Psi}_\omega, \, \mathbf{\Psi}_z]_S,$$

whereas, by (9.33) and (9.34),

$$\lim_{\alpha_i \to \beta} \mathbf{\Psi}_{\alpha_i}(\zeta) = \begin{pmatrix} C_1 \\ C_2 \end{pmatrix} G(\zeta)^{-1} \quad (\zeta \in \mathbb{C} \setminus \{\beta\}).$$

Thus, by Fatou's lemma,

$$\begin{aligned}
P_S &= \left[ \begin{pmatrix} C_1 \\ C_2 \end{pmatrix} G(\zeta)^{-1}, \begin{pmatrix} C_1 \\ C_2 \end{pmatrix} G(\zeta)^{-1} \right]_S \\
&= \frac{1}{2\pi} \int_0^{2\pi} \lim_{\alpha_i \to \beta} \mathbf{\Psi}_{\alpha_i}(e^{it})^* \begin{pmatrix} I_p & -S(e^{it}) \\ -S(e^{it})^* & I_q \end{pmatrix} \mathbf{\Psi}_{\alpha_i}(e^{it}) dt \\
&\leq \lim_{\alpha_i \to \beta} \frac{1}{2\pi} \int_0^{2\pi} \mathbf{\Psi}_{\alpha_i}(e^{it})^* \begin{pmatrix} I_p & -S(e^{it}) \\ -S(e^{it})^* & I_q \end{pmatrix} \mathbf{\Psi}_{\alpha_i}(e^{it}) dt \\
&= \lim_{\alpha_i \to \beta} [\mathbf{\Psi}_{\alpha_i}, \mathbf{\Psi}_{\alpha_i}]_S = \lim_{\alpha_i \to \beta} \mathbf{L}_{\alpha_i}(\alpha_i) = P_{\mathbf{L}}. \qquad (9.54)
\end{aligned}$$

**Step 2.** *The nontangential limits* (9.22) *and* (9.29) *(for* $j = 0, \cdots, m$) *and*

$$\angle \lim_{z \to \beta} \mathbf{\Psi}_z(\zeta) = \begin{pmatrix} C_1 \\ C_2 \end{pmatrix} G(\zeta)^{-1} = C G(\zeta)^{-1} \qquad (9.55)$$

*all exist for every point* $\zeta \in \mathbb{C} \setminus \{\beta\}$.

**Proof of Step 2:** In view of Lemma 1.3 and the bound (9.53), $S$ satisfies conditions (1.13) and (1.14): i.e., the functions $B$ and $\widetilde{B}$ defined in (1.13) and (1.14), belong to $\mathbf{H}_2^{p \times n}$ and $\mathbf{H}_2^{n \times q}$, respectively. Thus, by Corollary 7.6,

$$\angle \lim_{z \to \beta} (z - \beta) B(z) = 0 \quad \text{and} \quad \angle \lim_{z \to \beta} (z - \beta) \widetilde{B}(z) = 0,$$

which serve to establish (9.30) and (9.24), respectively. Therefore, the nontangential limits (9.22) and (9.15) exist by Lemma 9.3, the limits (9.29) and (9.16) exist by Lemma 9.4 and, finally, the limit (9.55) exists by Lemma 9.5.

**Step 3.** *The matrices* $C_1$ *and* $C_2$ *are subject to* (9.17).

## 9. A MULTIPLE ANALOGUE OF THE CARATHÉODORY–JULIA THEOREM

**Proof of Step 3:** On the one hand, by Leibnitz's rule and formulas (9.9) and (9.22), it follows that

$$\frac{1}{m!} \angle \lim_{z \to \beta} \frac{d^m}{dz^m} \left( \widetilde{H}(z) C_1^* S(z) C_2 \widetilde{G}(z) \right)$$

$$= \angle \lim_{z \to \beta} \sum_{j=0}^{m} \frac{1}{j!(m-j)!} \left( \widetilde{H}(z) C_1^* S(z) \right)^{(j)} C_2 \widetilde{G}^{(m-j)}(z)$$

$$= -\sum_{j=0}^{m} (T^*)^{m-j} C_2^* C_2 \left( -\beta T \right)^j N^{-j-1}.$$

On the other hand, in view of (9.9) and (9.29),

$$\frac{1}{m!} \angle \lim_{z \to \beta} \frac{d^m}{dz^m} \left( \widetilde{H}(z) C_1^* S(z) C_2 \widetilde{G}(z) \right)$$

$$= \angle \lim_{z \to \beta} \sum_{j=0}^{m} \frac{1}{j!(m-j)!} \widetilde{H}^{(j)}(z) C_1^* \left( S(z) C_2 \widetilde{G}(z) \right)^{(m-j)}$$

$$= -\sum_{j=0}^{m} (T^*)^{m-j} C_1^* C_1 \left( -\beta T \right)^j N^{-j-1}.$$

Upon comparison of the expressions on the right hand sides in the last two sequences of equalities we get (9.17).

**Step 4.** *The columns of the mvf $B(\zeta) = (I_p, -S(\zeta)) CG(\zeta)^{-1}$ belong to the space $\mathcal{H}(S)$ and*

$$\langle By, Bx \rangle_{\mathcal{H}(S)} = \angle \lim_{z \to \beta} x^* \mathbf{V}(z) y \quad (\forall \, x, y \in \mathbb{C}^n), \tag{9.56}$$

*where $\mathbf{V}(z)$ is the mvf defined in (9.11).*

**Proof of Step 4:** Let $B_z(\zeta)$ be the mvf defined by (9.36). By (9.49), there exists a sequence of points $\gamma_i \in \mathcal{U}_\beta$ tending to $\beta$ such that $B_{\gamma_i} x$ tends weakly to a limit $g \in \mathcal{H}(S)$. Since weak convergence implies pointwise convergence in $\mathbb{D}$, via the reproducing kernel,

$$g(\zeta) = \lim_{\gamma_i \to \beta} B_{\gamma_i}(\zeta) x \in \mathcal{H}(S) \tag{9.57}$$

for every point $\zeta \in \mathbb{D}$. Therefore, as $B_z(\zeta) = (I_p, -S(\zeta)) \mathbf{\Psi}_z(\zeta)$, the following nontangential limit exists by (9.55):

$$\angle \lim_{z \to \beta} B_z(\zeta) = \angle \lim_{z \to \beta} (I_p, -S(\zeta)) \mathbf{\Psi}_z(\zeta) = (I_p, -S(\zeta)) CG(\zeta)^{-1} = B(\zeta), \tag{9.58}$$

and, together with (9.57), implies that

$$B(\zeta) x = g(\zeta) \in \mathcal{H}(S).$$

Moreover, by (9.36),

$$\langle By, B_z x \rangle_{\mathcal{H}(S)} = \frac{1}{m!} \left\langle By, \frac{\partial^m}{\partial \bar{z}^m} \left( \Lambda_z C_1 \widetilde{H}(z)^* \right) x \right\rangle_{\mathcal{H}(S)}$$

$$= \frac{1}{m!} \frac{d^m}{dz^m} \left( x^* \widetilde{H}(z) C_1^* B(z) y \right) = x^* \mathbf{V}(z) y$$

and therefore, by the weak convergence,
$$\langle By,\ Bx\rangle_{\mathcal{H}(S)} = \angle\lim_{z\to\beta}\langle By,\ B_z x\rangle_{\mathcal{H}(S)},$$
which serves to establish (9.56).

**Step 5.** *The nontangential limit $P_{\mathbf{V}}$ in (9.18) exists. It is positive semidefinite and*
$$\langle By,\ Bx\rangle_{\mathcal{H}(S)} = x^* P_{\mathbf{V}} y \qquad (\forall\ x,\ y \in \mathbb{C}^n).$$

**Proof of Step 5:** This is immediate from (9.56).

**Step 6.** *The nontangential limit (9.13) exists and is equal to the limit (9.18).*

**Proof of Step 6:** By Remark 1.2, the matrix $P_S$ is positive semidefinite and satisfies the Lyapunov–Stein equation (1.16). Therefore, $S$ is a solution of the **aBIP**$(I_n, N, P_S, C)$. By Theorem 3.8, the corresponding mvf $\mathbf{W}$ defined by (3.12) belongs to the Carathéodory class $\mathcal{C}^{n\times n}$. Moreover, the identity (3.13) holds (with $M = I_n$ and $P = P_S$), by Lemma 3.7. Multiplying both of sides of (3.13) by $\widetilde{H}(z)$ on the left and by $\frac{\widetilde{H}(\omega)^*}{\rho_\omega(z)}$ on the right and taking advantage of (9.7) we get

$$\widetilde{H}(z)C_1^*\Lambda_\omega(z)C_1\widetilde{H}(\omega)^* = (z-\beta)^{m+1}(\bar\omega - \bar\beta)^{m+1}\frac{\mathbf{W}(z)+\mathbf{W}(\omega)^* - \widetilde{B}(z)\widetilde{B}(\omega)^*}{\rho_\omega(z)}$$
$$+\widetilde{H}(z)\left(C_1^* B(z) + B(\omega)^* C_1 - P_S\right)\widetilde{H}(\omega)^*.$$

Upon applying the operator $\dfrac{1}{(m!)^2}\dfrac{\partial^{2m}}{\partial z^m \partial\bar\omega^m}$ to both of sides of the last identity and taking into account (9.11) and the formula $\widetilde{H}^{(m)}(z) = m!I_n$, we get

$$\mathbf{L}_\omega(z) = \frac{1}{(m!)^2}\frac{\partial^{2m}}{\partial z^m \partial\bar\omega^m}\left((z-\beta)^{m+1}\frac{\mathbf{W}(z)+\mathbf{W}(\omega)^* - \widetilde{B}(z)\widetilde{B}(\omega)^*}{\rho_\omega(z)}(\bar\omega-\bar\beta)^{m+1}\right)$$
$$+\mathbf{V}(z) + \mathbf{V}(\omega)^* - P_S. \qquad (9.59)$$

Since $\mathbf{W} \in \mathcal{C}^{n\times n}$ and $\widetilde{B} \in \mathbf{H}_2^{n\times q}$, Lemmas 7.3 and 7.7 imply, that the limit of the first term on the right hand side in (9.59) exists as $z$ and $\omega$ tend to $\beta$ nontangentially and is positive semidefinite. The nontangential limits of the second and of the third terms exist by Step 5 and are equal to $P_{\mathbf{V}}$ by definition. Thus, the nontangential limit (9.13) exists and by (9.52), it is equal to $P_{\mathbf{L}}$. This completes the proof of implication (1) $\Rightarrow$ (2). Moreover, it now follows from (9.59) that
$$P_{\mathbf{L}} \geq 2P_{\mathbf{V}} - P_S.$$

Finally, upon taking advantage of (9.36) and (9.58), we get

$$P_{\mathbf{L}} = \angle\lim_{z,\omega\to\beta}\frac{1}{(m!)^2}\frac{\partial^{2m}}{\partial z^m \partial\bar\omega^m}\left(\widetilde{H}(z)C_1^*\Lambda_\omega(z)C_1\widetilde{H}(\omega)^*\right)$$
$$= \angle\lim_{z\to\beta}\frac{1}{m!}\frac{d^m}{dz^m}\left(\widetilde{H}(z)C_1^*\angle\lim_{\omega\to\beta}\frac{1}{m!}\frac{\partial^m}{\partial\bar\omega^m}\left(\Lambda_\omega(z)C_1\widetilde{H}(\omega)^*\right)\right)$$
$$= \angle\lim_{z\to\beta}\frac{1}{m!}\frac{d^m}{dz^m}\left(\widetilde{H}(z)C_1^*\angle\lim_{\omega\to\beta} B_\omega(z)\right)$$
$$= \angle\lim_{z\to\beta}\frac{1}{m!}\frac{d^m}{dz^m}\left(\widetilde{H}(z)C_1^* B(z)\right) = P_{\mathbf{V}}.$$

Thus,
$$P_S \geq P_{\mathbf{L}} = P_{\mathbf{V}},$$
which, together with (9.54), implies (9.19) and completes the proof. □

To get a complete analogue of Theorem 8.1 we have to clarify which statements in (9.13) and (9.15)–(9.18) are independent (i.e., imply all the others). The next two theorems establish different sets of implications.

THEOREM 9.8. *Let $S \in \mathcal{S}^{p \times q}$, $C_1 \in \mathbb{C}^{p \times n}$, $C_2 \in \mathbb{C}^{q \times n}$, $\beta \in \mathbb{T}$ and let $\widetilde{H}$ and $\widetilde{G}$ be given by (9.7) and (9.8), respectively. Suppose that the nontangential limits (9.15), (9.16) and (9.18) exist and that the last limit defines a Hermitian matrix $P_{\mathbf{V}}$. Then the nontangential limit $P_{\mathbf{L}}$ defined by (9.13) exists and $P_{\mathbf{L}} = P_{\mathbf{V}} = P_S$.*

The following example[4] shows that the theorem is false if $P_{\mathbf{V}}$ is not assumed to be Hermitian.

EXAMPLE 9.9. Let $S(z) = \dfrac{1+z}{2}$, $\beta = 1$, $m = 1$, $C_1 = (1, 0)$ and $C_2 = (1, \frac{1}{2})$. Then
$$\widetilde{H}(z) = \begin{pmatrix} z-1 & 0 \\ 1 & z-1 \end{pmatrix}, \quad \widetilde{G}(z) = \begin{pmatrix} 1-z & z \\ 0 & 1-z \end{pmatrix},$$
$$\lim_{z \to 1} \frac{d}{dz}\left(\widetilde{H}(z) C_1^* S(z)\right) = \frac{1}{2} \cdot \lim_{z \to 1} \frac{d}{dz}\begin{pmatrix} z^2 - 1 \\ 1+z \end{pmatrix} = \begin{pmatrix} 1 \\ \frac{1}{2} \end{pmatrix} = C_2^*$$

and
$$\lim_{z \to 1} \frac{d}{dz}\left(S(z) C_2 \widetilde{G}(z)\right) = \lim_{z \to 1} \frac{d}{dz}\left(\frac{1-z^2}{2}, \frac{(1+z)^2}{4}\right)$$
$$= -(1, 0) \begin{pmatrix} 1 & -1 \\ 0 & 1 \end{pmatrix} = -C_1 N^{-1}.$$

Furthermore, in view of (9.11),
$$P_{\mathbf{V}} = \lim_{z \to 1} \mathbf{V}(z) = \frac{1}{4} \cdot \lim_{z \to 1} \frac{d}{dz} \begin{pmatrix} 2(z-1) & 1-z \\ 2 & -1 \end{pmatrix} = \begin{pmatrix} \frac{1}{2} & -\frac{1}{4} \\ 0 & 0 \end{pmatrix}.$$

Thus, all the three limits in (9.15), (9.16) and (9.18) exist. However, $P_{\mathbf{V}}$ is not Hermitian and
$$\mathbf{L}_\omega(z) = \frac{\partial^2}{\partial z \partial \bar{\omega}}\left(\begin{pmatrix} z-1 \\ 1 \end{pmatrix} \frac{1 - \frac{1+z}{2} \cdot \frac{1+\bar{\omega}}{2}}{1 - z\bar{\omega}} (\bar{\omega} - 1, 1)\right)$$
$$= \begin{pmatrix} * & * \\ * & \dfrac{(1-z)(1-\bar{\omega})}{(1-z\bar{\omega})^3} \end{pmatrix}$$
is not bounded near the point $\beta = 1$.

**Proof of Theorem 9.8:** By Theorem 9.1, it suffices to check that (9.12) is in force for every point $z$ in a nontangential neighborhood $\mathcal{U}_\beta$ of $\beta$. We break the proof into steps.

---
[4]We wish to thank A. Kheifets for calling our attention to this example.

**Step 1.** *The nontangential limits*
$$\angle \lim_{z \to \beta} \frac{1}{j!} \frac{d^j}{dz^j} \left( \widetilde{H}(z) C_1^* B(z) \right) = (T^*)^{m-j} P_{\mathbf{V}} \tag{9.60}$$
*exist for* $j = 0, \ldots, m$.

**Proof of Step 1:** The proof is similar to the proof of the implication $(1) \Rightarrow (2)$ in Lemma 9.3. For $j = m$ the limit in (9.60) coincides with (9.18), which exists and is equal to $P_{\mathbf{V}}$ by assumption. Therefore, by Leibnitz's rule,
$$\angle \lim_{z \to \beta} \frac{d^m}{dz^m} \left( (z - \beta) \widetilde{H}(z) C_1^* B(z) \right) = m \angle \lim_{z \to \beta} \frac{d^{m-1}}{dz^{m-1}} \left( \widetilde{H}(z) C_1^* B(z) \right). \tag{9.61}$$
On the other hand, taking advantage of formula (9.26), we obtain
$$\angle \lim_{z \to \beta} \frac{d^m}{dz^m} \left( (z - \beta) \widetilde{H}(z) C_1^* B(z) \right)$$
$$= m! T^* P_{\mathbf{V}} + \angle \lim_{z \to \beta} \frac{d^m}{dz^m} \left( (z - \beta)^{m+1} C_1^* B(z) \right)$$
$$= m! T^* P_{\mathbf{V}} + \angle \lim_{z \to \beta} \frac{d^m}{dz^m} \left( C_1^* \left( C_1 - S(z) C_2 \right) \widetilde{G}(z) \right). \tag{9.62}$$
But, in view of (9.8) and (9.16), the last limit in (9.62) is easily seen to be equal to zero. Thus, upon comparing (9.61) and (9.62), we obtain the formula
$$\angle \lim_{z \to \beta} \frac{d^{m-1}}{dz^{m-1}} \left( \widetilde{H}(z) C_1^* B(z) \right) = (m-1)! T^* P_{\mathbf{V}}, \tag{9.63}$$
which proves (9.60) for $j = m - 1$. Much the same arguments may be used to verify (9.60) recursively for $j = m - 2, \ldots, 1, 0$.

**Step 2.** *The matrix $P_{\mathbf{V}}$ is a solution of the Stein equation*
$$P_{\mathbf{V}} - N^* P_{\mathbf{V}} N = C_1^* C_1 - C_2^* C_2. \tag{9.64}$$

**Proof of Step 2:** It is readily checked that
$$N^* \widetilde{H}(z) = z \widetilde{H}(z) - (z - \beta)^{m+1} I_n \quad \text{and} \quad G(z)^{-1} N = z^{-1} \left( G(z)^{-1} - I_n \right) \tag{9.65}$$
and hence, that
$$\widetilde{H}(z) C_1^* \left( C_1 - S(z) C_2 \right) G(z)^{-1} - N^* \widetilde{H}(z) C_1^* \left( C_1 - S(z) C_2 \right) G(z)^{-1} N$$
$$= \widetilde{H}(z) C_1^* \left( C_1 - S(z) C_2 \right) + (z - \beta)^{m+1} C_1^* \left( C_1 - S(z) C_2 \right) G(z)^{-1} N$$
$$= \widetilde{H}(z) C_1^* \left( C_1 - S(z) C_2 \right) + C_1^* \left( C_1 - S(z) C_2 \right) \widetilde{G}(z) N.$$
Thus, upon applying $\dfrac{1}{m!} \dfrac{d^m}{dz^m}$ to both sides of the last identity and invoking (9.15), (9.16), (9.18) and the identities
$$\widetilde{H}^{(m)}(\beta) = m! I_n \quad \text{and} \quad \widetilde{G}^{(m)}(\beta) = -m! N^{-1},$$
we obtain (9.64):
$$P_{\mathbf{V}} - N^* P_{\mathbf{V}} N = \frac{1}{m!} \angle \lim_{z \to \beta} \frac{d^m}{dz^m} \left( \widetilde{H}(z) C_1^* C_1 - \widetilde{H}(z) C_1^* S(z) C_2 \right.$$
$$\left. - C_1^* S(z) C_2 \widetilde{G}(z) N + C_1^* C_1 \widetilde{G}(z) N \right)$$
$$= C_1^* C_1 - C_2^* C_2.$$

## 9. A MULTIPLE ANALOGUE OF THE CARATHÉODORY–JULIA THEOREM

**Step 3.** Let the mvf $\widetilde{W}(z)$ be defined by the formula
$$\widetilde{W}(z) = -\widetilde{H}(z)P_{\mathbf{V}} + \widetilde{H}(z)C_1^* B(z). \tag{9.66}$$

Then the nontangential limit
$$\angle \lim_{z,\omega \to \beta} \frac{\partial^m}{\partial z^m}\left(\widetilde{W}(z)\left(\frac{\rho_\beta(z)}{\rho_\omega(z)}\right)^{m+1}\right) = 0. \tag{9.67}$$

**Proof of Step 3:** It follows readily from (9.9) and (9.60) that
$$\angle \lim_{z \to \beta} \widetilde{W}^{(j)}(z) = 0 \qquad (j = 0, \ldots, m)$$
and thus, by Lemma 7.8,
$$\lim_{z \to \beta} (z-\beta)^{-m} \widetilde{W}(z) = 0.$$

Let $U_\beta(\phi) \subset U_\beta(\widetilde{\phi})$ be Stoltz angles with vertices at $\beta$ which are defined as in (7.1). Choose a disk $\mathbb{D}_\beta(\delta)$ centered at $\beta$ with radius $\delta$ such that
$$\|(\zeta-\beta)^{-m}\widetilde{W}(\zeta)\| < \varepsilon \quad \text{for all} \quad \zeta \in U_\beta(\widetilde{\phi}) \cap \mathbb{D}_\beta(\delta).$$

Let $\widetilde{\psi}$ be an angle such that $0 < \widetilde{\psi} < \frac{\pi}{2} - \widetilde{\phi}$ and let $V_\beta(\psi)$ be the cone with vertex at the origin which is defined via (7.4). By (7.6), the inequality
$$\left|\frac{\rho_\beta(\zeta)}{\rho_\omega(\zeta)}\right| < \frac{|1-\zeta\bar{\beta}|}{1-|\zeta|} < 1 + \frac{1}{\cos(\widetilde{\phi}+\widetilde{\psi})}$$

holds for every $\zeta \in U_\beta(\widetilde{\phi}) \cap V_\beta(\widetilde{\psi})$ and for every $\omega \in \mathbb{D}$. Next, take a pair of points $z$ and $\omega$ in the nontangential neighborhood
$$\mathcal{U}_\beta := U_\beta(\phi) \cap V_\beta(\widetilde{\psi}) \cap \mathbb{D}_\beta(\delta/2)$$
of $\beta$ and let $\mathbb{T}_z$ be the maximal circle which is centered at $z$ and lies in $U_\beta(\widetilde{\phi})$. Then $\mathbb{T}_z$ also belongs to $\mathbb{D}_\beta(\delta)$ and therefore, the estimate (7.18) holds for every point $\zeta \in \mathbb{T}_z$. By (7.19),
$$\left|\frac{\zeta-\beta}{\zeta-z}\right| < 1 + \frac{1}{\sin(\widetilde{\phi}-\phi)} \quad (\forall\, \zeta \in \mathbb{T}_z).$$

Therefore, by Cauchy's formula,
$$\begin{aligned}
\left\|\frac{\partial^m}{\partial z^m}\left(\widetilde{W}(z)\frac{\rho_\beta(z)^{m+1}}{\rho_\omega(z)^{m+1}}\right)\right\| &= \left\|\frac{m!}{2\pi}\int_{\mathbb{T}_z}\frac{\widetilde{W}(\zeta)}{(\zeta-z)^{m+1}}\frac{\rho_\beta(\zeta)^{m+1}}{\rho_\omega(\zeta)^{m+1}}d\zeta\right\| \\
&= \left\|\frac{m!}{2\pi}\int_{\mathbb{T}_z}\frac{\widetilde{W}(\zeta)}{(\zeta-\beta)^m}\left(\frac{\zeta-\beta}{\zeta-z}\right)^m\left(\frac{\rho_\beta(\zeta)}{\rho_\omega(\zeta)}\right)^{m+1}\frac{d\zeta}{\zeta-z}\right\| \\
&\leq m!\max_{\zeta\in\mathbb{T}_z}\left\{\frac{\|\widetilde{W}(\zeta)\|}{|\zeta-\beta|^m}\left|\frac{\zeta-\beta}{\zeta-z}\right|^m\left|\frac{\rho_\beta(\zeta)}{\rho_\omega(\zeta)}\right|^{m+1}\right\} \\
&< m!\varepsilon\left(1+\frac{1}{\sin(\widetilde{\phi}-\phi)}\right)^m\left(1+\frac{1}{\cos(\widetilde{\phi}+\widetilde{\psi})}\right)^{m+1},
\end{aligned}$$

which proves (9.67).

**Step 4.** *The uniform estimate (9.12) is in force for every point $z$ in a nontangential neighborhood $\mathcal{U}_\beta$ of $\beta$.*

**Proof of Step 4:** Since $P_\mathbf{V}$ is Hermitian by assumption, we can apply Lemma 3.6 to conclude that the identity (3.11) holds with $M$, $P$ and $W$ replaced by $I_n$, $P_\mathbf{V}$ and $\widetilde{W}$, respectively:

$$C_1^* \left(I_p - S(z)S(\omega)^*\right) C_1 = H(z)\widetilde{W}(z)G(z) + G(\omega)^*\widetilde{W}(\omega)^* H(\omega)^*$$
$$+ \rho_w(z) P_\mathbf{V} + H(z) P_\mathbf{V} H(\omega)^*$$
$$- H(z)\widetilde{B}(z)\widetilde{B}(\omega)^* H(\omega)^*.$$

Multiplying both of sides of the latter identity by $\widetilde{H}(z)$ on the left and by $\frac{\widetilde{H}(\omega)^*}{\rho_\omega(z)}$ on the right and taking advantage of (3.10), (9.7) and (9.66), we get

$$\widetilde{H}(z)C_1^* \Lambda_\omega(z) C_1 \widetilde{H}(\omega)^* = \widetilde{W}(z)G(z)\frac{\widetilde{H}(\omega)^*}{\rho_\omega(z)} + \frac{\widetilde{H}(z)}{\rho_\omega(z)} G(\omega)^* \widetilde{W}(\omega)^*$$
$$+ \widetilde{H}(z) P_\mathbf{V} \widetilde{H}(\omega)^* + \frac{(z-\beta)^{m+1}(\bar{\omega}-\bar{\beta})^{m+1}}{\rho_\omega(z)} P_\mathbf{V}$$
$$- \frac{(z-\beta)^{m+1}(\bar{\omega}-\bar{\beta})^{m+1} \widetilde{B}(z)\widetilde{B}(\omega)^*}{\rho_\omega(z)}.$$

Applying the operator $\dfrac{1}{(m!)^2} \dfrac{\partial^{2m}}{\partial z^m \partial \bar{\omega}^m}$ to both sides of the last identity and invoking (9.20), we obtain

$$\mathbf{L}_\omega(z) + \frac{1}{(m!)^2} \frac{\partial^{2m}}{\partial z^m \partial \bar{\omega}^m} \left( \frac{(z-\beta)^{m+1}(\bar{\omega}-\bar{\beta})^{m+1} \widetilde{B}(z)\widetilde{B}(\omega)^*}{\rho_\omega(z)} \right)$$
$$= \frac{1}{m!} \frac{\partial^m}{\partial z^m} \left( \widetilde{W}(z) \left( \frac{\rho_\beta(z)}{\rho_\omega(z)} \right)^{m+1} \right) + \frac{1}{m!} \frac{\partial^m}{\partial \bar{\omega}^m} \left( \widetilde{W}(\omega)^* \left( \frac{\rho_\omega(\beta)}{\rho_\omega(z)} \right)^{m+1} \right)$$
$$+ P_\mathbf{V} + \frac{1}{(m!)^2} \frac{\partial^{2m}}{\partial z^m \partial \bar{\omega}^m} \left( \frac{(z-\beta)^{m+1}(\bar{\omega}-\bar{\beta})^{m+1}}{\rho_\omega(z)} P_\mathbf{V} \right). \quad (9.68)$$

By Step 3, the first two terms on the right hand side of (9.68) are bounded in some nontangential neighborhood $\mathcal{U}_\beta$ of $\beta$, whereas the last term is bounded by Lemma 7.7. Thus, upon letting $z = \omega$ and noting that both of terms on the left hand side of (9.68) are positive semidefinite in this setting, we may conclude that $\mathbf{L}_z(z)$ is uniformly bounded in $\mathcal{U}_\beta$. □

COROLLARY 9.10. *Let the nontangential limits in (9.15) and (9.13) exist and define the matrices $C_2$ and $P_\mathbf{L}$, respectively. Then $P_\mathbf{L}$ satisfies the Stein equation (8.4).*

PROOF. The existence of the nontangential limit (9.13) guarantees that Theorem 9.1 is applicable and hence that $P_\mathbf{L} = P_S$. Therefore, since $P_S$ is a solution of the Stein equation (8.4) by Remark 1.2, $P_\mathbf{L}$ is also a solution of (8.4). □

We conclude the section with the following modification of Theorem 9.8 which will be useful.

THEOREM 9.11. *Let $S \in \mathcal{S}^{p \times q}$, $C_1 \in \mathbb{C}^{p \times n}$, $C_2 \in \mathbb{C}^{q \times n}$ and let the nontangential limits (9.16) (or (9.15)) and (9.18) exist. Suppose further that the matrix $P_\mathbf{V} \in \mathbb{C}^{n \times n}$ defined via (9.18) is a Hermitian solution of the Stein equation (9.64). Then the nontangential limit $P_\mathbf{L}$ in (9.13) exists and $P_\mathbf{L} = P_\mathbf{V} = P_S$.*

PROOF. The proof is based on the proof of Theorem 9.8. The difference between the two formulations is that whereas in Theorem 9.8 we assumed that the tangential limits (9.15), (9.16) and (9.18) exist and that $P_\mathbf{V}$ is Hermitian, we now drop one of the two conditions (9.15), (9.16), but add (9.64). However, since the role of (9.15) in the proof of Theorem 9.8 was limited to establishing (9.64) and we are now assuming (9.64), it is clear that the conclusions of Theorem 9.8 prevail even if we drop (9.15).

Suppose next that (9.15), (9.18) and (9.64) are in force. By the proof of Theorem 9.8, it suffices to show that relations (9.60) hold for $j = 0, \ldots, m$. Upon denoting the limits in (9.61) by $m!X$ and making use of (9.15), (9.18) and the formula
$$(z - \beta)G(z)^{-1} = -z\beta G(z)^{-1}T - \beta I_n$$
we get
$$\begin{aligned}
X &= -\frac{\beta}{m!} \angle \lim_{z \to \beta} \frac{d^m}{dz^m} \left( \widetilde{H}(z)C_1^* \left(C_1 - S(z)C_2\right) \left(zG(z)^{-1}T + I_n\right) \right) \\
&= -\frac{\beta}{m!} \angle \lim_{z \to \beta} \frac{d^m}{dz^m} \left( z\widetilde{H}(z)C_1^* B(z)T \right) - \beta(C_1^* C_1 - C_2^* C_2) \\
&= -\beta^2 P_\mathbf{V} T - \beta X T - \beta(C_1^* C_1 - C_2^* C_2)
\end{aligned}$$
Since $P_\mathbf{V}$ satisfies (9.64), this is equivalent to
$$XN = -\beta P_\mathbf{V} T - P_\mathbf{V} + N^* P_\mathbf{V} N.$$
Therefore,
$$X = -P_\mathbf{V}(I + \beta T)N^{-1} + N^* P_\mathbf{V} = -\beta P_\mathbf{V} + N^* P_\mathbf{V} = T^* P_\mathbf{V}.$$
This proves (9.63), which coincides with (9.60) when $j = m - 1$. The recursive verification of (9.60) for $j = m - 2, \ldots, 1, 0$ relies on much the same arguments. □

## 10. On the solvability of a Stein equation

Let $r$ be an integer satisfying $1 \leq r \leq \min(p, q)$, let $n = r(m+1)$, let $M$, $N$, $T$ and $C$ be the matrices defined in (9.1)–(9.3). In this section we establish necessary and sufficient conditions for the Stein equation (8.4) based on this choice of $M$, $N$ and $C$ to have a solution.

LEMMA 10.1. *The upper triangular block matrix*
$$\mathbf{D} = \begin{pmatrix} \beta I_r & -\beta^2 I_r & \beta^3 I_r & \cdots & (-1)^m \beta^{m+1} I_r \\ 0 & -\beta^3 I_r & 2\beta^4 I_r & \cdots & (-1)^m \binom{m}{1} \beta^{m+2} I_r \\ \vdots & & \beta^5 I_r & \cdots & (-1)^m \binom{m}{2} \beta^{m+3} I_r \\ \vdots & & & \ddots & \vdots \\ 0 & \cdots & \cdots & 0 & (-1)^m \beta^{2m+1} I_r \end{pmatrix} \qquad (10.1)$$

*with the $r \times r$ block entries*
$$D_{j\ell} = \begin{cases} 0, & \text{if } j > \ell \\ (-1)^\ell \binom{\ell}{j} \beta^{\ell+j+1} I_r, & \text{if } j \leq \ell \end{cases} \qquad (10.2)$$

*is the unique solution of the Stein equation*

$$\mathbf{D} + \beta\left(I_n + \beta T^*\right)\mathbf{D}T = \beta E^* E, \tag{10.3}$$

*where $N$, $T$ and $E$ are the matrices defined by (9.1), (9.3) and (9.42), respectively.*

PROOF. Since $T^{m+1} = 0$, the Stein equation (10.3) has a unique solution, which is given by the formula

$$\mathbf{D} = \sum_{j=0}^{m} (-1)^j \beta^{j+1} \left(I_n + \beta T^*\right)^j E^* E T^j. \tag{10.4}$$

To show that the expression on the right hand side of (10.4) represents the matrix defined by (10.2), it suffices to rewrite (10.4) as

$$\begin{aligned}
\mathbf{D} &= \sum_{\ell=0}^{m} (-1)^\ell \beta^{\ell+1} \left( \sum_{j=0}^{\ell} \binom{\ell}{j} (\beta T^*)^j \right) E^* E T^\ell \\
&= \sum_{\ell=0}^{m} \sum_{j=0}^{\ell} (-1)^\ell \binom{\ell}{j} \beta^{\ell+j+1} T^{*j} E^* E T^\ell \\
&= \sum_{\ell=0}^{m} \sum_{j=0}^{\ell} D_{j\ell} T^{*j} E^* E T^\ell,
\end{aligned}$$

since all the block entries of the matrix $T^{*j} E^* E T^\ell$ are equal to zero except for the $j\ell$-th block which is equal to $I_r$. □

COROLLARY 10.2. *The matrix $\mathbf{D}$ defined via (10.2) also admits the representation*

$$\mathbf{D} = \sum_{j=0}^{m} T^{*j} E^* E (-\beta T)^j N^{-j-1}. \tag{10.5}$$

PROOF. It follows from (10.3) that

$$\mathbf{D}T = E^* E - \bar{\beta}\mathbf{D} - \beta T^* \mathbf{D}T$$

and therefore,

$$\mathbf{D}N = \mathbf{D}(\bar{\beta} I_n + T) = E^* E - \beta T^* \mathbf{D}T.$$

Thus, $\mathbf{D}$ satisfies the Stein equation

$$\mathbf{D} + \beta T^* \mathbf{D}T N^{-1} = E^* E N^{-1}, \tag{10.6}$$

which has a unique solution given by the formula (10.5). □

Besides $C_1$ and $C_2$, it will be useful to consider the upper triangular block Toeplitz matrices

$$\mathbf{C}_1 = \begin{pmatrix} \xi_0 & \xi_1 & \cdots & \xi_m \\ 0 & \xi_0 & \ddots & \vdots \\ \vdots & & \ddots & \xi_1 \\ 0 & \cdots & 0 & \xi_0 \end{pmatrix} \quad \text{and} \quad \mathbf{C}_2 = \begin{pmatrix} \eta_0 & \eta_1 & \cdots & \eta_m \\ 0 & \eta_0 & \ddots & \vdots \\ \vdots & & \ddots & \eta_1 \\ 0 & \cdots & 0 & \eta_0 \end{pmatrix}. \tag{10.7}$$

It will be convenient to express these block matrices in the form

$$\mathbf{C}_k = \sum_{j=0}^{m} T^{*j} E^* C_k T^j \quad (k = 1, 2). \tag{10.8}$$

REMARK 10.3. In formula (10.8), the $T$ on the left of $E^* C_k$ is different from the $T$ on the right of $E^* C_k$. Strictly speaking we should introduce three block shift matrices $T_p$, $T_q$, $T_r$ and write

$$\mathbf{C}_1 = \sum_{j=0}^{m} T_p^{*j} F_p^* C_1 T_r^j \quad \text{and} \quad \mathbf{C}_2 = \sum_{j=0}^{m} T_q^{*j} E_q^* C_2 T_r^j.$$

However, in order to keep the formulas simple, we have not indicated this in the notation, but rely instead on the context. We shall usually maintain this abuse of notation for the matrices $N$, $\mathbf{D}$, $E$ and $\mathbf{U}$ (see formula (10.14) below) also. Occasionally we shall write $\mathbf{C}_k$ "commutes" with $T$, with quotation marks to emphasize the fact that the $T$ on the left is not required to have the same block size as the $T$ on the right.

The following result will be useful.

PROPOSITION 10.4. *Let* $E$, $T$, $C_1$, $C_2$, $\mathbf{C}_1$, $\mathbf{C}_2$ *be the matrices given by* (9.42), (9.3), (9.2), (10.8) *and let*

$$\mathbf{E}(z) = \sum_{j=0}^{m} \frac{E T^j}{(z-\beta)^{j+1}} = \left( \frac{I_r}{z-\beta}, \frac{I_r}{(z-\beta)^2}, \ldots, \frac{I_r}{(z-\beta)^{m+1}} \right). \tag{10.9}$$

*Then*

$$E G(z)^{-1} = -\mathbf{E}(z)\mathbf{D} \quad \text{and} \quad C_k G(z)^{-1} = -\mathbf{E}(z)\mathbf{D}\mathbf{C}_k \quad (k = 1, 2). \tag{10.10}$$

PROOF. On account of (10.2),

$$\frac{z^\ell}{(1-z\bar{\beta})^{\ell+1}} = \frac{z^\ell \beta^{\ell+1}}{(\beta-z)^{\ell+1}} = (-1)^{\ell+1} \sum_{j=0}^{\ell} \binom{\ell}{j} \frac{\beta^{\ell+j+1}}{(z-\beta)^{j+1}} = -\sum_{j=0}^{\ell} \frac{D_{j\ell}}{(z-\beta)^{j+1}},$$

which together with (9.4), (9.42) and (10.9) implies the first relation in (10.10). Making use of this relation and of

$$C_k T^j = E T^j \mathbf{C}_k \quad \text{for} \quad j \geq 0 \quad \text{and} \quad k = 1, 2, \tag{10.11}$$

we obtain

$$C_k G(z)^{-1} = E G(z)^{-1} \mathbf{C}_k = -\mathbf{E}(z)\mathbf{D}\mathbf{C}_k \quad (k = 1, 2)$$

and complete the proof. □

THEOREM 10.5. *Let* $C_1 \in \mathbb{C}^{p \times n}$, $C_2 \in \mathbb{C}^{q \times n}$, *let* $\mathbf{C}_1$ *and* $\mathbf{C}_2$ *be defined via* (10.8) *and let* $\widetilde{H}$ *and* $G^{-1}$ *be the mvf's defined by* (9.7) *and* (9.4), *respectively. Then the following statements are equivalent:*

(1) $C_1$ *and* $C_2$ *are subject to* (9.17).
(2) *The mvf*

$$\widetilde{H}(z) \{C_1^* C_1 - C_2^* C_2\} G(z)^{-1} \tag{10.12}$$

*is a matrix polynomial.*

(3) $\mathbf{C}_1$ and $\mathbf{C}_2$ are subject to

$$\mathbf{C}_1^* \mathbf{U}_p \mathbf{D}_p \mathbf{C}_1 = \mathbf{C}_2^* \mathbf{U}_q \mathbf{D}_q \mathbf{C}_2, \tag{10.13}$$

where $\mathbf{D}_r$ is defined by (10.1) and

$$\mathbf{U}_k = \begin{pmatrix} 0 & \cdots & 0 & I_k \\ \vdots & \cdots & I_k & 0 \\ 0 & \cdots & \cdots & \vdots \\ I_k & 0 & \cdots & 0 \end{pmatrix}. \tag{10.14}$$

(4) The bottom block rows in formula (10.13) match, i.e.,

$$(\xi_0^*, \ldots, \xi_m^*)\mathbf{D}\mathbf{C}_1 = (\eta_0^*, \ldots, \eta_m^*)\mathbf{D}\mathbf{C}_2. \tag{10.15}$$

(5) The Stein equation (8.4) has at least one solution $P$ (in which case the Hermitian matrix $\frac{1}{2}(P + P^*)$ is also a solution).

If any one of the above statements holds true, then the degree of the matrix polynomial (10.12) does not exceed $m - 1$.

PROOF. (1) $\Leftrightarrow$ (3). Making use of the equalities (10.11) and

$$(T^*)^{m-j} E^* = \mathbf{U} T^{*j} E^* \quad (j = 0, \ldots m; \ k = 1, 2), \tag{10.16}$$

which follow readily from the definitions (9.3), (9.42), and (10.14) of the matrices $T$, $E$ and $\mathbf{U}$, respectively, and taking into account that $\mathbf{C}_k$ intertwines $N$, because of the upper triangular block Toeplitz structure, we get

$$\sum_{j=0}^{m} (T^*)^{m-j} C_k^* C_k (-\beta T)^j N^{-j-1} = \sum_{j=0}^{m} \mathbf{C}_k^* (T^*)^{m-j} E^* E (-\beta T)^j N^{-j-1} \mathbf{C}_k$$

$$= \mathbf{C}_k^* \mathbf{U} \left( \sum_{j=0}^{m} (T^*)^j E^* E (-\beta T)^j N^{-j-1} \right) \mathbf{C}_k$$

$$= \mathbf{C}_k^* \mathbf{U} \mathbf{D} \mathbf{C}_k \quad (k = 1, 2).$$

This proves the desired equivalence.

(2) $\Leftrightarrow$ (3). Upon taking advantage of the expansions (9.6) and (9.7) and using (10.11), we get

$$\widetilde{H}(z) C_k^* C_k G(z)^{-1}$$

$$= - \left( \sum_{i=0}^{m} (z - \beta)^{m-i} (T^*)^i \right) C_k^* C_k \left( \sum_{j=0}^{m} (z - \beta)^{-j-1} (-\beta T)^j N^{-j-1} \right)$$

$$= -\mathbf{C}_k^* \sum_{i,j=0}^{m} (z - \beta)^{m-i-j-1} (T^*)^i E^* E (-\beta T)^j N^{-j-1} \mathbf{C}_k$$

$$= -\mathbf{C}_k^* \left( p(z) + q(z) \right) \mathbf{C}_k \quad (k = 1, 2),$$

where

$$p(z) = \sum_{\ell=0}^{m-1} (z - \beta)^\ell \sum_{j=0}^{m-\ell-1} (T^*)^{m-\ell-j-1} E^* E (-\beta T)^j N^{-j-1}$$

and
$$q(z) = \sum_{\ell=1}^{m+1}(z-\beta)^{-\ell}\sum_{j=0}^{m}(T^*)^{m+\ell-j-1}E^*E(-\beta T)^j N^{-j-1}.$$

Using (10.16) and (10.5) we obtain
$$\begin{aligned}q(z) &= \sum_{\ell=0}^{m}(z-\beta)^{-\ell-1}(T^*)^{\ell}\sum_{j=0}^{m}(T^*)^{m-j}E^*E(-\beta T)^j N^{-j-1}\\ &= \sum_{\ell=0}^{m}(z-\beta)^{-\ell-1}(T^*)^{\ell}\mathbf{U}\sum_{j=0}^{m}(T^*)^j E^*E(-\beta T)^j N^{-j-1}\\ &= \sum_{\ell=0}^{m}(z-\beta)^{-\ell-1}(T^*)^{\ell}\mathbf{UD}.\end{aligned}$$

Therefore,
$$\widetilde{H}(z)\left\{C_1^*C_1 - C_2^*C_2\right\}G(z)^{-1} = \mathbf{C}_2^*\left(p(z)+q(z)\right)\mathbf{C}_2 - \mathbf{C}_1^*\left(p(z)+q(z)\right)\mathbf{C}_1$$

is a matrix polynomial if and only if
$$\mathbf{C}_1^* q(z)\mathbf{C}_1 = \mathbf{C}_2^* q(z)\mathbf{C}_2,$$

which in turn, is equivalent to the equalities
$$\mathbf{C}_1^*(T^*)^{\ell}\mathbf{UDC}_1 = \mathbf{C}_2^*(T^*)^{\ell}\mathbf{UDC}_2 \quad \text{for } \ell=0,\ldots,m.$$

But these equalities are equivalent to (10.13), since $\mathbf{C}_k$ "commutes" with $T$.

(3) $\Leftrightarrow$ (4). The implication (3) $\Rightarrow$ (4) is selfevident. To obtain the converse, we need to show that equality of the bottom block rows in the matrices $\mathbf{C}_1^*\mathbf{U}_p\mathbf{D}_p\mathbf{C}_1$ and $\mathbf{C}_2^*\mathbf{U}_q\mathbf{D}_q\mathbf{C}_2$ implies the "whole" equality (10.13).

Upon multiplying both sides of (10.6) by $T^j$ on the left and taking into account the fact that $T^j E^* = 0$ for $j \geq 1$ and that $TT^*\mathbf{D}T = \mathbf{D}T$, we obtain
$$T^j\mathbf{D} = -\beta T^{j-1}\mathbf{D}TN^{-1} \quad (j=1,\ldots,m),$$

which implies recursively that
$$T^j\mathbf{D} = \mathbf{D}(-\beta TN^{-1})^j \quad \text{for } j=1,\ldots,m. \tag{10.17}$$

Thus, since the matrix $-\beta TN^{-1}$ "commutes" with $\mathbf{C}_1$ and $\mathbf{C}_2$,
$$(\xi_0^*, \ldots, \xi_m^*)\mathbf{DC}_1(-\beta TN^{-1})^j = (\xi_0^*, \ldots, \xi_m^*)T^j\mathbf{DC}_1 = (0, \xi_0^*, \ldots, \xi_j^*)\mathbf{DC}_1$$

and
$$(\eta_0^*, \ldots, \eta_m^*)\mathbf{DC}_2(-\beta TN^{-1})^j = (\eta_0^*, \ldots, \eta_m^*)T^j\mathbf{DC}_2 = (0, \eta_0^*, \ldots, \eta_j^*)\mathbf{DC}_2.$$

Therefore, (10.15) implies that
$$(0, \xi_0^*, \ldots, \xi_j^*)\mathbf{DC}_1 = (0, \eta_0^*, \ldots, \eta_j^*)\mathbf{DC}_2 \quad \text{for } j=0,\ldots,m,$$

which expresses the equality of the $j$-th block rows in (10.13).

(1) $\Rightarrow$ (5). Let (9.17) be in force. Then, since
$$ET^k(T^*)^{k-j} = ET^j \quad (0 \leq j \leq k \text{ and } k=0,\ldots,m), \tag{10.18}$$

it follows readily, upon multiplying (9.17) by $ET^m$ on the left and by $N$ on the right, that

$$E \sum_{j=0}^{m} T^j \left(C_2^* C_2 - C_1^* C_1\right) (-\beta T)^j N^{-j} = 0.$$

Therefore,

$$E \left(C_2^* C_2 - C_1^* C_1\right) = \beta E \left\{ \sum_{j=1}^{m} T^j \left(C_2^* C_2 - C_1^* C_1\right) (-\beta T)^{j-1} N^{-j} \right\} T. \tag{10.19}$$

Let $P_0$ denote the term inside the curly brackets on the right, i.e.,

$$P_0 := \sum_{j=0}^{m-1} T^{j+1} \left(C_2^* C_2 - C_1^* C_1\right) (-\beta T)^j N^{-j-1}. \tag{10.20}$$

We now show that the matrix $P_0$ satisfies (8.4). To do this we multiply (10.19) by $E^*$ on the left and invoke the identity

$$E^* E = I_n - T^* T, \tag{10.21}$$

to conclude that

$$\beta P_0 T + C_1^* C_1 - C_2^* C_2 = T^* T \left(\beta P_0 T + C_1^* C_1 - C_2^* C_2\right). \tag{10.22}$$

The last equality exhibits the fact that the top block row of the matrix indicated on the left is equal to zero. On the other hand, it is readily seen from the definition (10.20) of $P_0$ that it satisfies the Stein equation

$$P_0 + \beta T P_0 T N^{-1} = T \left(C_2^* C_2 - C_1^* C_1\right) N^{-1},$$

and hence, upon multiplying through by $T^*$ on the left and by $N$ on the right, we obtain

$$T^* P_0 N = T^* T \left(C_2^* C_2 - C_1^* C_1 - \beta P_0 T\right) = C_2^* C_2 - C_1^* C_1 - \beta P_0 T,$$

in view of (10.22). Thus,

$$C_2^* C_2 - C_1^* C_1 = T^* P_0 N + \beta P_0 T,$$

which is equivalent to (8.4), since

$$T^* P_0 N + \beta P_0 T = (N^* - \bar{\beta} I_n) P_0 N + \beta P_0 \left(N - \bar{\beta} I_n\right) = N^* P_0 N - P_0.$$

(5) $\Rightarrow$ (2). Let $P \in \mathbb{C}^{n \times n}$ satisfy (8.4). Then, making use of (9.65), we get

$$\begin{aligned} \widetilde{H}(z) \{C_1^* C_1 - C_2^* C_2\} G(z)^{-1} &= \widetilde{H}(z) \{P - N^* P N\} G(z)^{-1} \\ &= \widetilde{H}(z) P + P \widetilde{G}(z) N, \end{aligned} \tag{10.23}$$

which is obviously a matrix polynomial of degree not more than $m$. By (9.8) and (9.7), the leading terms of $\widetilde{G}$ and $\widetilde{H}$ are $-z^m N^{-1}$ and $z^m I_n$, respectively, and thus, the expression on the right hand side of (10.23) is a polynomial of degree not more than $m-1$. □

LEMMA 10.6. *All the solutions of the homogeneous Stein equation*

$$\widehat{P} - N^* \widehat{P} N = 0 \tag{10.24}$$

## 10. ON THE SOLVABILITY OF A STEIN EQUATION

*are given by the formula*

$$\widehat{P}(L) = \mathbf{U} \sum_{j=0}^{m} T^{*j} E^* L (-\beta T)^j N^{-j}, \qquad (10.25)$$

*where* $\mathbf{U}$ *is given by* (10.14) *and* $L$, *the lowest block row of* $\widehat{P}(L)$, *is a parameter varying over* $\mathbb{C}^{r \times n}$.
*Moreover,* $\widehat{P}(L)$ *is positive semidefinite if and only if all its entries are equal to zero except the bottom right hand* $r \times r$ *block, which is positive semidefinite or, equivalently, if and only if* $L$ *is of the form*

$$L = (0, \ldots, 0, \alpha), \quad \text{where} \quad 0 \leq \alpha \in \mathbb{C}^{r \times r}.$$

PROOF. First we show that every matrix $\widehat{P}(L)$ of the form (10.25) is a solution of the homogeneous Stein equation (10.24). Since $T^{m+1} = 0$ and $T$ and $N$ commute, it follows readily from (10.25) that

$$
\begin{aligned}
E\widehat{P}(L)T &= -\bar{\beta} E \mathbf{U} \sum_{j=0}^{m} T^{*j} E^* L (-\beta T)^{j+1} N^{-j} \\
&= -\bar{\beta} E \mathbf{U} \sum_{j=0}^{m-1} T^{*j} E^* L (-\beta T)^{j+1} N^{-j} = 0,
\end{aligned}
$$

because $E \mathbf{U} T^{*j} E^* = 0$ for $j = 0, \ldots, m-1$. Therefore, in view of (10.21),

$$(I - T^*T) \widehat{P}(L) T = 0. \qquad (10.26)$$

Next, since $\mathbf{U}^2 = I_n$, it is readily seen from (10.25) that the matrix $\mathbf{U}\widehat{P}(L)$ is a solution of the equation

$$\mathbf{U}\widehat{P}(L) + \beta T^* \mathbf{U} \widehat{P}(L) T N^{-1} = E^* L.$$

But this is equivalent to the equation

$$\widehat{P}(L) N + \beta T \widehat{P}(L) T = \mathbf{U} E^* L N,$$

since $\mathbf{U} T^* \mathbf{U} = T$. Multiplying the last equation by $T^*$ on the left and taking advantage of (10.26), we conclude that $\widehat{P}(L)$ is a solution of the equation

$$\beta \widehat{P} T + T^* \widehat{P} N = 0,$$

which in turn, is equivalent to (10.24). Thus, every matrix of the form (10.25) is a solution of (10.24).
Now let $Q$ be any solution of (10.24), let $A$ denote its lowest block row and let $\widehat{P}(A)$ be the matrix defined via (10.25). Then the matrix $R := \widehat{P}(A) - Q$ satisfies the Stein equation (10.24) and has zero entries in the lowest block row, that is

$$R - N^* R N = 0 \quad \text{and} \quad E T^m R = 0. \qquad (10.27)$$

Upon writing the first of these equalities in the form

$$\beta R T + T^* R N = 0,$$

and then multiplying it by $ET^m$ on the left and invoking (10.18) and the second equality in (10.27), we conclude that $ET^{m-1}R = 0$. In much the same way we obtain recursively that $ET^j R = 0$ for all $j = 0, \ldots, m$ and hence that

$$R = \sum_{j=0}^{m} T^{*j} E^* E T^j R = 0.$$

Thus, $Q = \widehat{P}(A)$, which means that every solution of (10.24) is of the form (10.25).

Furthermore, substituting the block decomposition

$$\widehat{P}(L) = \left(\widehat{P}_{ij}\right)_{i,j=0}^{m}, \qquad \widehat{P}_{ij} \in \mathbb{C}^{r \times r}$$

into (10.24) and comparing all the $r \times r$ blocks we get

$$\widehat{P}_{0j} = \widehat{P}_{j0} = 0 \quad \text{and} \quad \beta \widehat{P}_{i,j+1} + \bar{\beta} \widehat{P}_{i+1,j} + \widehat{P}_{ij} = 0 \quad (i, j = 0, \ldots, m-1).$$

Therefore, since $\widehat{P} \geq 0$, we also have $\widehat{P}_{0m} = \widehat{P}_{m0} = 0$. Moreover, since $\beta \widehat{P}_{11} + \bar{\beta} \widehat{P}_{20} + \widehat{P}_{10} = 0$, we now get $\widehat{P}_{11} = 0$. Since $\widehat{P} \geq 0$, it follows that $\widehat{P}_{1j} = \widehat{P}_{j1} = 0$ for all $j = 0, \ldots, m$. In much the same manner we obtain recursively that all the block entries $\widehat{P}_{ij}$ are equal to zero except for the block $\widehat{P}_{mm} = \alpha$, which is positive semidefinite, since $\widehat{P}$ is. □

COROLLARY 10.7. *Let (9.17) be in force. Then the matrix $P_0$ defined by (10.20) is the only solution of the Stein equation (8.4) which has zero entries in the lowest block row. Moreover, the set of all the solutions of the Stein equation (8.4) is parametrized by the formula*

$$P = P_0 + \widehat{P}(L),$$

*where $\widehat{P}(L)$ is defined in (10.25) and $L$, the lowest block row of $\widehat{P}(L)$, is a parameter varying over $\mathbb{C}^{r \times n}$. Thus, for each $L \in \mathbb{C}^{r \times n}$ there exists exactly one solution $P$ of the Stein equation (8.4) whose the lowest block row coincides with $L$.*

COROLLARY 10.8. *Let $0 \leq \widehat{P} \leq P$. Then the following statements are equivalent:*

(1) *Both $\widehat{P}$ and $P$ are solutions of the Stein equation (8.4).*
(2) *All the entries of $P$ are equal to the corresponding entries of $\widehat{P}$ except for the bottom right hand $r \times r$ blocks which satisfy the condition $\widehat{P}_{mm} \leq P_{mm}$.*

For the proof, note that (1) is equivalent to the fact that the matrix $\check{P} = P - \widetilde{P}$ is a positive semidefinite solution of the homogeneous Stein equation (10.24) and then apply Lemma 10.6.

The following formulas will prove useful in the sequel and give yet another indication of the significance of the condition (10.15).

LEMMA 10.9. *Let $N$ and $C$ be as in (9.1) and (9.2), let $\mathbf{H}_\eta$ and $\mathbf{H}_\xi$ be the Hankel matrices defined by*

$$\mathbf{H}_\xi = \begin{pmatrix} \xi_1 & \xi_2 & \cdots & \xi_{m+1} \\ \xi_2 & \xi_3 & \cdots & \xi_{m+2} \\ \vdots & \vdots & & \vdots \\ \xi_{m+1} & \xi_{m+2} & \cdots & \xi_{2m+1} \end{pmatrix} \qquad (10.28)$$

and
$$\mathbf{H}_\eta = \begin{pmatrix} \eta_1 & \eta_2 & \cdots & \eta_{m+1} \\ \eta_2 & \eta_3 & \cdots & \eta_{m+2} \\ \vdots & \vdots & & \vdots \\ \eta_{m+1} & \eta_{m+2} & \cdots & \eta_{2m+1} \end{pmatrix} \quad (10.29)$$

for some choice of $\xi_{m+1}, \ldots, \xi_{2m+1} \in \mathbb{C}^{p \times r}$ and $\eta_{m+1}, \ldots, \eta_{2m+1} \in \mathbb{C}^{q \times r}$ and let

$$P_\xi = \mathbf{H}_\xi^* \mathbf{D} \mathbf{C}_1 \quad \text{and} \quad P_\eta = \mathbf{H}_\eta^* \mathbf{D} \mathbf{C}_2, \quad (10.30)$$

where $\mathbf{C}_j$ and $\mathbf{D}$ are the matrices given in (10.7) and (10.1), respectively. Then:

(1) The matrices $P_\xi$ and $P_\eta$ satisfy the identities

$$P_\xi - N^* P_\xi N = -C_1^* C_1 + E^* \begin{pmatrix} \xi_0^* & \cdots & \xi_m^* \end{pmatrix} \mathbf{D} \mathbf{C}_1 N \quad (10.31)$$

and

$$P_\eta - N^* P_\eta N = -C_2^* C_2 + E^* \begin{pmatrix} \eta_0^* & \cdots & \eta_m^* \end{pmatrix} \mathbf{D} \mathbf{C}_2 N, \quad (10.32)$$

respectively.

(2) The matrix $P_\eta - P_\xi$ is a solution of the Stein equation (8.4) if and only if

$$\begin{pmatrix} \eta_0^* & \cdots & \eta_m^* \end{pmatrix} \mathbf{D} \mathbf{C}_2 = \begin{pmatrix} \xi_0^* & \cdots & \xi_m^* \end{pmatrix} \mathbf{D} \mathbf{C}_1. \quad (10.33)$$

PROOF. Upon multiplying both sides of (10.6) by $T$ on the left and by $N$ on the right we obtain

$$TDN + \beta TT^* \mathbf{D} T = 0,$$

and, since $\mathbf{D}$ is upper block triangular and hence $(I - TT^*)\mathbf{D}T = 0$, we see that

$$TDN = -\beta DT.$$

Moreover, since $N = \bar{\beta} I + T$ and both of the matrices $N$ and $T$ commute with $\mathbf{C}_1$, it follows from formula (10.30) that

$$\begin{aligned} N^* P_\xi N - P_\xi &= T^* P_\xi N + \beta P_\xi T \\ &= T^* \mathbf{H}_\xi^* \mathbf{D} \mathbf{C}_1 N + \beta \mathbf{H}_\xi^* \mathbf{D} \mathbf{C}_1 T \\ &= T^* \mathbf{H}_\xi^* \mathbf{D} N \mathbf{C}_1 + \beta \mathbf{H}_\xi^* \mathbf{D} T \mathbf{C}_1 \\ &= T^* \mathbf{H}_\xi^* \mathbf{D} N \mathbf{C}_1 - \mathbf{H}_\xi^* T \mathbf{D} N \mathbf{C}_1 \\ &= \left( T^* \mathbf{H}_\xi^* - \mathbf{H}_\xi^* T \right) \mathbf{D} N \mathbf{C}_1. \end{aligned}$$

It is readily seen that, due to the Hankel structure (10.28) of $\mathbf{H}_\xi^*$,

$$T^* \mathbf{H}_\xi^* - \mathbf{H}_\xi^* T = \begin{pmatrix} 0 \\ \xi_1^* \\ \vdots \\ \xi_m^* \end{pmatrix} E - E^* \begin{pmatrix} 0 & \xi_1^* & \cdots & \xi_m^* \end{pmatrix}$$

and thus,

$$P_\xi - N^* P_\xi N = E^* \begin{pmatrix} 0 & \xi_1^* & \cdots & \xi_m^* \end{pmatrix} \mathbf{D} N \mathbf{C}_1 - \begin{pmatrix} 0 \\ \xi_1^* \\ \vdots \\ \xi_m^* \end{pmatrix} E \mathbf{D} N \mathbf{C}_1. \quad (10.34)$$

By (10.6),

$$E \mathbf{D} N \mathbf{C}_1 = E(E^* E - \beta T^* \mathbf{D} T) \mathbf{C}_1 = E \mathbf{C}_1 = C_1$$

and therefore,
$$\begin{pmatrix} 0 \\ \xi_1^* \\ \vdots \\ \xi_m^* \end{pmatrix} E\mathbf{DNC}_1 = \begin{pmatrix} 0 \\ \xi_1^* \\ \vdots \\ \xi_m^* \end{pmatrix} C_1 \qquad (10.35)$$

and
$$E^* \begin{pmatrix} \xi_0^* & 0 & \cdots & 0 \end{pmatrix} \mathbf{DNC}_1 = E^* \xi_0^* E \mathbf{DNC}_1 = E^* \xi_0^* C_1.$$

Next,
$$E^* \begin{pmatrix} 0 & \xi_1^* & \cdots & \xi_m^* \end{pmatrix} \mathbf{DNC}_1$$
$$= E^* \begin{pmatrix} \xi_0^* & \xi_1^* & \cdots & \xi_m^* \end{pmatrix} \mathbf{DNC}_1 - E^* \begin{pmatrix} \xi_0^* & 0 & \cdots & 0 \end{pmatrix} \mathbf{DNC}_1$$
$$= E^* \begin{pmatrix} \xi_0^* & \xi_1^* & \cdots & \xi_m^* \end{pmatrix} \mathbf{DNC}_1 - E^* \xi_0^* C_1. \qquad (10.36)$$

Upon substituting (10.35) and (10.36) into (10.34) and taking advantage of the equality
$$\begin{pmatrix} 0 \\ \xi_1^* \\ \vdots \\ \xi_m^* \end{pmatrix} + E^* \xi_0^* = C_1^*,$$

we arrive at (10.31). Identity (10.32) is verified in much the same way and the second statement of the lemma is an easy consequence of the first one. $\square$

Our next objective is to reexpress $P_0$ in terms of the strictly upper block triangular parts
$$\mathbf{H}_\xi^u = \begin{pmatrix} \xi_1 & \cdots & \xi_m & 0 \\ \vdots & \ddots & \vdots & \vdots \\ \xi_m & \cdots & 0 & 0 \\ 0 & \cdots & 0 & 0 \end{pmatrix} \quad \text{and} \quad \mathbf{H}_\eta^u = \begin{pmatrix} \eta_1 & \cdots & \eta_m & 0 \\ \vdots & \ddots & \vdots & \vdots \\ \eta_m & \cdots & 0 & 0 \\ 0 & \cdots & 0 & 0 \end{pmatrix}$$

of the Hankel matrices $\mathbf{H}_\xi$ and $\mathbf{H}_\eta$ (defined in (10.28) and (10.29), respectively) and then to express $\widehat{P}(L)$ in terms of the corresponding lower block triangular parts
$$\mathbf{H}_\xi^\ell = \mathbf{H}_\xi - \mathbf{H}_\xi^u \quad \text{and} \quad \mathbf{H}_\eta^\ell = \mathbf{H}_\eta - \mathbf{H}_\eta^u. \qquad (10.37)$$

LEMMA 10.10. *Let $N$ and $C$ be as in (9.1) and (9.2). Then the special solution $P_0$ of the Stein equation (8.4) that is defined by formula (10.20) can also be written as*
$$P_0 = (\mathbf{H}_\eta^u)^* \mathbf{DC}_2 - (\mathbf{H}_\xi^u)^* \mathbf{DC}_1. \qquad (10.38)$$

PROOF. The proof rests essentially on the observation that $TC_2^*, \ldots, T^{m+1}C_2^*$ are the block columns of the matrix $(\mathbf{H}_\eta^u)^*$:
$$(\mathbf{H}_\xi^u)^* = \begin{pmatrix} TC_1^*, & T^2 C_1^*, & \ldots, & T^{m+1} C_1^* \end{pmatrix}$$

and similarly,
$$(\mathbf{H}_\eta^u)^* = \begin{pmatrix} TC_2^*, & T^2 C_2^*, & \ldots, & T^{m+1} C_2^* \end{pmatrix}.$$

On the other hand, since $E\mathbf{D} = EN^{-1}$ (as follows immediately from (10.6)), we conclude by (10.17), that

$$\begin{aligned} ET^j\mathbf{D}\mathbf{C}_k &= E\mathbf{D}(-\beta TN^{-1})^j\mathbf{C}_k \\ &= EN^{-1}(-\beta TN^{-1})^j\mathbf{C}_k \\ &= E\mathbf{C}_k(-\beta T)^j N^{-j-1} \\ &= C_k(-\beta T)^j N^{-j-1} \quad (k=1,2;\ j=0,\ldots,m), \end{aligned}$$

which expresses the fact that $C_k(-\beta T)^j N^{-j-1}$ is the $j$-th block row of the matrix $\mathbf{D}\mathbf{C}_k$ for $k=1,2$. Thus,

$$\sum_{j=0}^{m-1} T^{j+1} C_1^* C_1 (-\beta T)^j N^{-j-1} = (\mathbf{H}_\xi^u)^* \mathbf{D}\mathbf{C}_1,$$

$$\sum_{j=0}^{m-1} T^{j+1} C_2^* C_2 (-\beta T)^j N^{-j-1} = (\mathbf{H}_\eta^u)^* \mathbf{D}\mathbf{C}_2$$

and the difference yields (10.38), in view of (10.20). □

LEMMA 10.11. *Let $\mathbf{C}_j$ and $\mathbf{D}$ be defined by formulas (10.7) and (10.1), respectively, let $\mathbf{H}_\xi^\ell$ and $\mathbf{H}_\eta^\ell$ be the Hankel matrices defined by (10.37) for some $\xi_{m+1},\ldots,\xi_{2m+1} \in \mathbb{C}^{p\times r}$ and $\eta_{m+1},\ldots,\eta_{2m+1} \in \mathbb{C}^{q\times r}$ and let the matrix $\widehat{P}(L)$ be defined by (10.25).*

(1) *If $L = (\xi_{m+1}^*, \ldots, \xi_{2m+1}^*)\mathbf{D}\mathbf{C}_1$, then $\widehat{P}(L) = (\mathbf{H}_\xi^\ell)^*\mathbf{D}\mathbf{C}_1$.*
(2) *If $L = (\eta_{m+1}^*, \ldots, \eta_{2m+1}^*)\mathbf{D}\mathbf{C}_2$, then $\widehat{P}(L) = (\mathbf{H}_\eta^\ell)^*\mathbf{D}\mathbf{C}_2$.*
(3) *If*

$$L = (\eta_{m+1}^*, \ldots, \eta_{2m+1}^*)\mathbf{D}\mathbf{C}_2 - (\xi_{m+1}^*, \ldots, \xi_{2m+1}^*)\mathbf{D}\mathbf{C}_1,$$

*then*

$$\widehat{P}(L) = (\mathbf{H}_\eta^\ell)^*\mathbf{D}\mathbf{C}_2 - (\mathbf{H}_\xi^\ell)^*\mathbf{D}\mathbf{C}_1.$$

PROOF. If $L$ is as in the first statement, then, by (10.25) and (10.17),

$$\begin{aligned} \widehat{P}(L) &= \mathbf{U}\sum_{j=0}^{m} T^{*j}E^* \left(\xi_{m+1}^*, \ldots, \xi_{2m+1}^*\right) \mathbf{D}\mathbf{C}_1(-\beta T)^j N^{-j} \\ &= \mathbf{U}\sum_{j=0}^{m} T^{*j}E^* \left(\xi_{m+1}^*, \ldots, \xi_{2m+1}^*\right) \mathbf{D}(-\beta T)^j N^{-j}\mathbf{C}_1 \\ &= \mathbf{U}\sum_{j=0}^{m} T^{*j}E^* \left(\xi_{m+1}^*, \ldots, \xi_{2m+1}^*\right) T^j\mathbf{D}\mathbf{C}_1 \\ &= \mathbf{U}\begin{pmatrix} \xi_{m+1}^* & \xi_{m+2}^* & \cdots & \xi_{2m+1}^* \\ 0 & \xi_{m+1}^* & \cdots & \xi_{2m}^* \\ \vdots & \ddots & \ddots & \vdots \\ 0 & \cdots & 0 & \xi_{m+1}^* \end{pmatrix}\mathbf{D}\mathbf{C}_1 = (\mathbf{H}_\xi^\ell)^*\mathbf{D}\mathbf{C}_1, \end{aligned}$$

which completes the proof of the first assertion. The proof of the second assertion is similar and the proof of the third is immediate from the first two and the fact that

$$\widehat{P}(L_1 - L_2) = \widehat{P}(L_1) - \widehat{P}(L_2).$$

LEMMA 10.12. *Let $\mathbf{C}_j$ and $\mathbf{D}$ be defined by formulas (10.7) and (10.1), respectively.*

(1) *If* $\operatorname{rank} \xi_0 = r$, *then every matrix* $L \in \mathbb{C}^{r \times n}$, $n = (m+1)r$, *can be expressed in the form* $L = \left(\xi_{m+1}^*, \ldots, \xi_{2m+1}^*\right) \mathbf{DC}_1$ *for some choice of* $\xi_{m+1}, \ldots, \xi_{2m+1} \in \mathbb{C}^{p \times r}$.

(2) *If* $\operatorname{rank} \eta_0 = r$, *then every matrix* $L \in \mathbb{C}^{r \times n}$, $n = (m+1)r$, *can be expressed in the form* $L = \left(\eta_{m+1}^*, \ldots, \eta_{2m+1}^*\right) \mathbf{DC}_2$ *for some choice of* $\eta_{m+1}, \ldots, \eta_{2m+1} \in \mathbb{C}^{q \times r}$.

PROOF. If $\operatorname{rank} \xi_0 = r$, then the matrix $\mathbf{C}_1$ is left invertible. Let $\mathbf{C}_1^{-L}$ denote its left inverse. Then clearly the choice $\left(\xi_{m+1}^*, \ldots, \xi_{2m+1}^*\right) = L\mathbf{C}_1^{-L}\mathbf{D}^{-1}$ satisfies the claim in the first assertion. The proof of the second assertion is similar. □

REMARK 10.13. If $M$, $N$, $C_1$ and $C_2$ are as in (9.1) and (9.2), but with $p = q = 1$, then, in view of Theorem 7.7 of [5], every solution $P$ of the Stein equation (8.4) is subject to the same Iohvidov law as a Hankel matrix. Consequently, if $P \neq 0$ is also positive semidefinite, then it must either be of the form

$$P = \begin{pmatrix} 0 & \cdots & 0 & 0 \\ \vdots & & \vdots & \vdots \\ 0 & \cdots & 0 & 0 \\ 0 & \cdots & 0 & \delta \end{pmatrix} \text{ for some } \delta > 0,$$

or of the form

$$P = \begin{pmatrix} P_{11} & P_{12} \\ P_{21} & P_{22} \end{pmatrix},$$

where $P_{11} > 0$ and

$$P_{22} - P_{21}P_{11}^{-1}P_{21} = \begin{pmatrix} 0 & \cdots & 0 & 0 \\ \vdots & & \vdots & \vdots \\ 0 & \cdots & 0 & 0 \\ 0 & \cdots & 0 & \delta \end{pmatrix} \text{ for some } \delta \geq 0,$$

since the Schur complement of $P_{11}$ is subject to the same Iohvidov law as $P$.

If $\xi_0 = \eta_0 = 1$ and $P = H_\eta^* \mathbf{DC}_2$ is a positive semidefinite solution of equation (8.4), then it will be of the first form if and only if $\eta_1 = 0$ (or, equivalently, if and only if $\eta_1 = \cdots = \eta_{2m} = 0$ and $\eta_{2m+1}^* = (-1)^m \beta^{2m+1}\delta$).

## 11. Positive definite solutions of the Stein equation

In this section we continue the analysis of the Stein equation (8.4) when $N$ is of the form (9.1) and $C$ is partitioned as in (9.2). We shall assume that

$$P = (P_{k\ell})_{k,\ell=0}^m, \qquad P_{k\ell} \in \mathbb{C}^{r \times r}, \tag{11.1}$$

is partitioned conformally with $N$. In addition to Theorem 10.5, where necessary and sufficient conditions for the solvability of the the Stein equation (8.4) have been established, we shall present necessary and sufficient conditions for this equation to have a positive definite solution and shall describe the set of all such solutions.

Substituting the block decompositions (9.1), (9.2) and (11.1) into (8.4) and comparing all the $r \times r$–blocks we get the system
$$\begin{aligned}\beta P_{0j} &= \eta_0^* \eta_{j+1} - \xi_0^* \xi_{j+1} & (j=0,\ldots,m-1) \\ \beta P_{i+1,j} + \bar{\beta} P_{i,j+1} + P_{ij} &= \eta_{i+1}^* \eta_{j+1} - \xi_{i+1}^* \xi_{j+1} & \\ & \quad (i=0,\ldots,m-1;\ j=1,\ldots,m-1),\end{aligned} \quad (11.2)$$
which, upon being solved recursively, leads to
$$P_{k\ell} = \sum_{i=0}^{k} \sum_{j=0}^{i} (-1)^i \binom{i}{j} \bar{\beta}^{i+j+1} \left( \eta_{k-i}^* \eta_{\ell+j+1} - \xi_{k-i}^* \xi_{\ell+j+1} \right) \quad (11.3)$$
for
$$0 \leq \ell + k \leq m - 1. \quad (11.4)$$
Thus, for every solution $P$ of the Stein equation (8.4), the entries with the indexes from the domain (11.4) are uniquely defined by the right hand side of the equation. Moreover, these entries satisfy the symmetry
$$P_{k\ell} = P_{\ell k}^* \quad (0 \leq \ell + k \leq m - 1). \quad (11.5)$$
This symmetry follows from the above mentioned uniqueness and the fact that $P^*$ satisfies the Stein equation (8.4) if and only if $P$ does.

Let $\mathbf{d}_j(P)$, $j = 0, \ldots, 2m$, denote the $j$-th "southwest-northeast" block diagonal of $P$. Then all the block entries $P_{st}$ in $\mathbf{d}_j(P)$ meet the condition $s + t = j$. In particular, with a selfevident interpretation,
$$\mathbf{d}_j(P) = (P_{j,0},\ P_{j-1,1},\ \ldots,\ P_{0,j}) \quad \text{for} \quad j = 0, \ldots, m-1$$
and
$$\mathbf{d}_j(P) = (P_{m,j-m},\ P_{m-1,j-m+1},\ \ldots,\ P_{j-m,m}) \quad \text{for} \quad j = m, \ldots, 2m.$$
It has been already mentioned that the diagonals $\mathbf{d}_0(P)$, ..., $\mathbf{d}_{m-1}(P)$ are uniquely defined by (8.4) and are Hermitian in the sense that (11.5) holds. It follows from (11.2) that for $j \geq m$, the diagonal $\mathbf{d}_j(P)$ is uniquely defined by any one entry $P_{j-i,i}$ and by the previous diagonal $\mathbf{d}_{j-1}(P)$. In particular, $\mathbf{d}_j(P)$ is uniquely defined by $P_{m,j-m}$ and $\mathbf{d}_{j-1}(P)$ for $j \geq m$.

The following lemma will enable us to construct a positive definite solution of the Stein equation (8.4) recursively.

LEMMA 11.1. *Let $C_1$ and $C_2$ be of the form (9.2) and satisfy (9.17), let $j$ be a fixed integer ($m - 1 \leq j \leq 2m - 1$) and let the blocks $P_{k\ell}$ of $P$ be given and satisfy*
$$\beta P_{k+1,\ell-1} + \bar{\beta} P_{k,\ell} + P_{k,\ell-1} = \eta_{k+1}^* \eta_\ell - \xi_{k+1}^* \xi_\ell \quad \text{and} \quad P_{k\ell} = P_{\ell k}^* \quad (11.6)$$
*for $0 \leq k + \ell \leq j$. Then there exist $r \times r$ matrices $P_{j+1,0}$, $P_{j,1}$, ..., $P_{0,j+1}$ such that relations (11.6) hold for $0 \leq k + \ell \leq j + 1$.*

*Moreover, if $j \geq m$ and $j = 2t$ is even, $P_{tt}$ can be chosen equal to any prescribed Hermitian $r \times r$ matrix $\gamma$, i.e., the matrices $P_{m,j-m}, \ldots, P_{j-m,m}$ can be chosen so that (11.6) holds and $P_{tt} = \gamma$.*

PROOF. It is clear from the recursion in (11.6) that if the $P_{i,j}$ are known for $0 \leq i + j \leq k$ and if $s, t$ are nonnegative integers with $s + t = k + 1$ (and of course $k + 1 \leq m$), then $P_{s,t}$ determines $P_{s-1,t+1}$, and vice versa. Thus, proceeding recursively, it is easy to obtain a solution of the Stein equation (8.4). But, if $P$ is a solution, then $P^*$ is also a solution. Therefore, $\frac{1}{2}(P + P^*)$ is a Hermitian solution. But this means that it is possible to chose the entries from the outset to meet the symmetry condition in (11.6) also. □

Let us consider the upper left $(\widehat{m} + 1) \times (\widehat{m} + 1)$ block submatrix

$$\Pi_{\widehat{m}} = (P_{k,\ell})_{k,\ell=0}^{\widehat{m}}, \quad \text{where} \quad \widehat{m} = \begin{cases} (m-2)/2 & \text{if } m \text{ is even} \\ (m-1)/2 & \text{if } m \text{ is odd} \end{cases} \quad (11.7)$$

of $P$. It is readily checked that $\Pi_{\widehat{m}}$ is the largest principle block submatrix of $P$ which sits strictly above the diagonal $\mathbf{d}_m(P)$ and hence is uniquely specified by the right hand side of equation (8.4). Moreover, $\Pi_{\widehat{m}}$ is Hermitian, by (11.5). Let $\widehat{\mathbf{C}}_1$, $\widehat{\mathbf{C}}_2$ and $\widehat{\mathbf{D}}$ be the matrices defined as in (10.7) and (10.1) but with $m$ replaced by $\widehat{m}$ and let

$$\widehat{\mathbf{H}}_\eta = \begin{pmatrix} \eta_1 & \eta_2 & \cdots & \eta_{\widehat{m}+1} \\ \eta_2 & \eta_3 & \cdots & \eta_{\widehat{m}+2} \\ \vdots & \vdots & & \vdots \\ \eta_{\widehat{m}+1} & \eta_{\widehat{m}+2} & \cdots & \eta_{2\widehat{m}+1} \end{pmatrix} \quad (11.8)$$

and

$$\widehat{\mathbf{H}}_\xi = \begin{pmatrix} \xi_1 & \xi_2 & \cdots & \xi_{\widehat{m}+1} \\ \xi_2 & \xi_3 & \cdots & \xi_{\widehat{m}+2} \\ \vdots & \vdots & & \vdots \\ \xi_{\widehat{m}+1} & \xi_{\widehat{m}+2} & \cdots & \xi_{2\widehat{m}+1} \end{pmatrix}. \quad (11.9)$$

It follows from (11.3) and (10.2) that

$$P_{k\ell} = \sum_{i=0}^{k} \sum_{j=0}^{i} (D_{ji})^* \left( \eta_{k-i}^* \eta_{\ell+j+1} - \xi_{k-i}^* \xi_{\ell+j+1} \right)$$

$$= (\eta_k^*, \eta_{k-1}^*, \ldots, \eta_0^*, 0) \widehat{\mathbf{D}}^* \begin{pmatrix} \eta_{\ell+1} \\ \eta_{\ell+2} \\ \vdots \\ \eta_{\ell+\widehat{m}+1} \end{pmatrix}$$

$$- (\xi_k^*, \xi_{k-1}^*, \ldots, \xi_0^*, 0) \widehat{\mathbf{D}}^* \begin{pmatrix} \xi_{\ell+1} \\ \xi_{\ell+2} \\ \vdots \\ \xi_{\ell+\widehat{m}+1} \end{pmatrix}$$

for $k, \ell = 0, \ldots, \widehat{m}$, which with help of (11.8) and (11.9) can be written in the matrix form as

$$\Pi_{\widehat{m}} = \widehat{\mathbf{C}}_2^* \widehat{\mathbf{D}}^* \widehat{\mathbf{H}}_\eta - \widehat{\mathbf{C}}_1^* \widehat{\mathbf{D}}^* \widehat{\mathbf{H}}_\xi$$

or, since $P$ is Hermitian,

$$\Pi_{\widehat{m}} = \widehat{\mathbf{H}}_\eta^* \widehat{\mathbf{D}} \widehat{\mathbf{C}}_2 - \widehat{\mathbf{H}}_\xi^* \widehat{\mathbf{D}} \widehat{\mathbf{C}}_1. \quad (11.10)$$

THEOREM 11.2. *Let $N$ be a matrix of the form (9.1). Then the Stein equation (8.4) has a positive definite solution if and only if:*
  (1) *The matrices $C_1$ and $C_2$ are subject to (9.17) and*
  (2) *The matrix $\Pi_{\widehat{m}}$ defined by (11.10) is positive definite.*

PROOF. The necessity part is clear: if the Stein equation (8.4) has a solution $P$, then condition (1) is fulfilled by Theorem 10.5. Moreover, if $P$ is positive definite, then the submatrix $\Pi_{\widehat{m}}$ (which is uniquely specified by (11.10)) is also positive definite.

Now let conditions (1) and (2) be in force. Then there exists a positive definite solution $P$ of the Stein equation (8.4). Indeed, using (11.3) we define the diagonals $\mathbf{d}_0(P), \ldots, \mathbf{d}_{m-1}(P)$ of $P$. The remaining $m+1$ diagonals are defined recursively to meet (11.6) with the added constraint that the entries $P_{tt}$ for $t$ running from $m$ to $2m$ are chosen to keep the matrix positive definite. This is easily seen to be doable by invoking the well known formulas for Schur complements, i.e., from the decomposition

$$\Pi_j = (P_{k\ell})_{k,\ell=0}^j = \begin{pmatrix} \Pi_{j-1} & Y_j^* \\ Y_j & P_{j,j} \end{pmatrix}, \quad \text{where} \quad Y_j = (P_{j,0}, P_{j,1}, \ldots, P_{j,j-1}) \tag{11.11}$$

and the evident factorization

$$\Pi_j = \begin{pmatrix} I & 0 \\ Y_j \Pi_{j-1}^{-1} & I \end{pmatrix} \begin{pmatrix} \Pi_{j-1} & 0 \\ 0 & P_{jj} - Y_j \Pi_{j-1}^{-1} Y_j^* \end{pmatrix} \begin{pmatrix} I & \Pi_{j-1}^{-1} Y_j^* \\ 0 & I \end{pmatrix},$$

which implies that, if $\Pi_{j-1} > 0$, then the matrix $\Pi_j$ is positive definite if and only if $P_{jj} > Y_j \Pi_{j-1}^{-1} Y_j^*$. □

Note that if condition (2) in Theorem 11.2 is relaxed to $\Pi_{\widehat{m}} \geq 0$, then there is no guarantee that the corresponding Stein equation has a positive semidefinite solution. Indeed, it is readily checked that the matrices

$$C_1 = (\xi_0, \xi_1) = \begin{pmatrix} 1 & 1 \\ 0 & 0 \end{pmatrix}, \quad C_2 = (\eta_0, \eta_1) = \begin{pmatrix} 0 & 1 \\ 1 & 1 \end{pmatrix} \quad \text{and} \quad N = \begin{pmatrix} \bar{\beta} & 1 \\ 0 & \bar{\beta} \end{pmatrix}$$

satisfy (9.17); $m = 1$ and $\widehat{m} = 0$. Furthermore, if $P$ satisfies (8.4), then

$$\Pi_{\widehat{m}} = P_{00} = \bar{\beta}(\eta_0^* \eta_1 - \xi_0^* \xi_1) = 0 \quad \text{and} \quad \beta P_{10} + \bar{\beta} P_{01} = \eta_1^* \eta_1 - \xi_1^* \xi_1 = 1. \tag{11.12}$$

Thus, if $P$ is positive semidefinite, the first relation in (11.12) implies $P_{10} = P_{01} = 0$, which contradicts the second equality in (11.12).

However, in conclusion, we mention the following analogue of Theorem 11.2, which is applicable in the degenerate case.

THEOREM 11.3. *Given matrix $N$ of the form (9.1), the Stein equation (8.4) has a positive semidefinite solution $P$ if and only if:*
  (1) *The matrices $C_1$ and $C_2$ are subject to (9.17).*
  (2) *The matrix $\Pi_{\widehat{m}}$ defined by (11.10) is positive semidefinite.*
  (3) *The matrices $P_{k\ell}$ defined by (11.3) for $0 \leq \ell + k \leq m-1$, are subject to*

$$\text{Ker } \Pi_j \subseteq \text{Ker } (P_{m-j-1,0}, P_{m-j-1,1}, \ldots, P_{m-j-1,j}) \quad (j = 0, \ldots, m - \widehat{m} - 2),$$

*where $\Pi_j$ is the submatrix of $P$ defined in (11.11) and $\widehat{m}$ is the integer given in (11.7).*

The necessity part is clear and follows from the nonnegativity of a solution $P$ of the Stein equation (8.4). Conversely, if conditions (1)–(3) are fulfilled, a positive semidefinite solution $P$ of (8.4) may be constructed recursively, as in the proof of Theorem 11.2.

## 12. A Carathéodory-Fejér boundary problem

This section is devoted to a Carathéodory-Fejér boundary problem that we shall refer to as the $\widehat{\mathbf{CFBP}}$. It is formulated below in terms of the matrix polynomials

$$\mathcal{A}_\ell(z) = \sum_{j=0}^{\ell}(z-\beta)^j \xi_j^* \quad \text{and} \quad \mathcal{B}_\ell(z) = \sum_{j=0}^{\ell}(z-\beta)^j \eta_j^* \quad (\ell = 0, \ldots, m) \qquad (12.1)$$

based on the components of the matrices $C_1$ and $C_2$ (the first of which were defined in terms of $C_1$ in (9.44)) and the $r \times r$ entries

$$\mathbf{L}_{m\ell}(z,\omega) = \frac{1}{(m!)^2}\frac{\partial^{2m}}{\partial z^m \partial \bar{\omega}^m}\left(\mathcal{A}_m(z)\Lambda_\omega(z)\mathcal{A}_\ell(\omega)^*(\bar{\omega}-\bar{\beta})^{m-\ell}\right) \qquad (12.2)$$

in the bottom block row of the kernel $\mathbf{L}_\omega(z)$ given in (9.10).

The $\widehat{\mathbf{CFBP}}$: Given $\xi_0, \ldots, \xi_m \in \mathbb{C}^{p \times r}$, $\eta_0, \ldots, \eta_m \in \mathbb{C}^{q \times r}$, $\alpha_0, \ldots, \alpha_m \in \mathbb{C}^{r \times r}$ and a point $\beta \in \mathbb{T}$, find necessary and sufficient conditions for the existence of a Schur function $S \in \mathcal{S}^{p \times q}$ such that:

$$\angle \lim_{z \to \beta}(z-\beta)^{-m}\{\mathcal{A}_m(z)S(z) - \mathcal{B}_m(z)\} = 0, \qquad (12.3)$$

$$\angle \lim_{z,\omega \to \beta}\mathbf{L}_{m\ell}(z,\omega) = \alpha_\ell \quad (0 \le \ell < m), \qquad (12.4)$$

and the nontangential limit $\angle \lim_{z,\omega \to \beta}\mathbf{L}_{mm}(z,w)$ exists and meets the constraint

$$\angle \lim_{z,\omega \to \beta}\mathbf{L}_{mm}(z,\omega) \le \alpha_m. \qquad (12.5)$$

REMARK 12.1. By Lemma 7.8, condition (12.3) is equivalent to

$$\frac{1}{\ell!}\angle \lim_{z \to \beta}\frac{d^\ell}{dz^\ell}\left(\mathcal{A}_m(z)S(z)\right) = \eta_\ell^* \quad (\ell = 0, \ldots, m). \qquad (12.6)$$

Moreover, condition (12.3) also is clearly equivalent to the set of conditions

$$\angle \lim_{z \to \beta}(z-\beta)^{-\ell}\{\mathcal{A}_\ell(z)S(z) - \mathcal{B}_\ell(z)\} = 0 \quad (\ell = 0, \ldots, m). \qquad (12.7)$$

Our next objective is to identify the $\widehat{\mathbf{CFBP}}$ with an appropriately defined $\widehat{\mathbf{aBIP}}$. The following preliminary result will be useful.

THEOREM 12.2. *Let $N$ and $C$ be as in (9.1) and (9.2), respectively, let $P$ be a positive semidefinite solution of the Stein equation (8.4) and let $S$ belong to $\mathcal{S}^{p \times q}$. Then the following statements are equivalent:*

(1) *$S$ belongs to $\widehat{\mathcal{S}}(I_n, N, P, C)$.*
(2) *The nontangential limit (9.13) exists and the value $P_\mathbf{L}$ of this limit is subject to the bounds $0 \le P_\mathbf{L} \le P$.*

PROOF. Let $S$ be a solution of $\widehat{\mathbf{aBIP}}(I_n, N, P, C)$. Then the functions $B$ and $\widetilde{B}$ defined by (1.13) and (1.14) belong to $\mathbf{H}_2^{p \times n}$ and $\mathbf{H}_2^{n \times q}$, respectively, and hence (9.15) and (9.16) hold by the proof of Step 2 of Theorem 9.1.

## 12. A CARATHÉODORY-FEJÉR BOUNDARY PROBLEM

Furthermore, since $S \in \widehat{\mathcal{S}}(I_n, N, P, C)$, the function $\mathbf{W}(z)$ defined by (3.12) belongs to the Carathéodory class $\mathcal{C}^{n \times n}$ and takes the value $\frac{1}{2}P$ at the origin. Therefore, it admits a Riesz–Herglotz representation of the form (8.17) and

$$P = 2\mathbf{W}(0) = \frac{1}{\pi} \int_0^{2\pi} d\sigma(t). \tag{12.8}$$

The mvf $W(z)$ defined by (3.10) satisfies

$$W(z) = \frac{\mathbf{W}(z) - \frac{1}{2}P}{z} = \frac{\mathbf{W}(z) - \mathbf{W}(0)}{z}$$

and hence, in view of Lemma 7.2,

$$\angle \lim_{z \to \beta} \left((z-\beta)^{m+1} W(z)\right)^{(m)} = -\frac{m!}{\pi} \sigma(\{t_0\}) \qquad (\beta = e^{it_0}). \tag{12.9}$$

It follows from (3.10) and (9.7) that

$$(z-\beta)^{m+1} W(z) = -\widetilde{H}(z) P + \widetilde{H}(z) C_1^* \left(C_1 - S(z) C_2\right) G(z)^{-1}.$$

Upon substituting the latter relation into (12.9) and making use of (9.11) and the first equality in (9.9), we get

$$-m! P + m! \angle \lim_{z \to \beta} \mathbf{V}(z) = -\frac{m!}{\pi} \sigma(\{t_0\}),$$

which implies that the limit in (9.18) exists and defines a Hermitian matrix $P_\mathbf{V}$ that is subject to the inequality $P_\mathbf{V} \leq P$, since $\sigma(\{t_0\}) \geq 0$. Moreover, in view of (12.8),

$$P_\mathbf{V} = \frac{1}{\pi} \left(\int_0^{2\pi} d\sigma(t) - \sigma(\{t_0\})\right) \geq 0.$$

Since relations (9.15), (9.16) and (9.18) are in force, the nontangential limit $P_\mathbf{L}$ defined by (9.13) exists and $P_\mathbf{L} = P_\mathbf{V} = P_S$, by Theorem 9.8.

Conversely, if (2) is in force, then Theorem 9.1 is applicable and guarantees that $P_S = P_\mathbf{L}$. Therefore, $P_S \leq P$ and hence, by Lemma 1.3, $S$ belongs to $\widehat{\mathcal{S}}(I_n, N, P, C)$. □

Some necessary conditions for the $\widehat{\mathbf{CFBP}}$ to be solvable are given in the next theorem.

THEOREM 12.3. *Let $S$ be a solution of the* $\widehat{\mathbf{CFBP}}$ *and let $N$, $C_1$ and $C_2$ be the matrices defined in (9.1) and (9.2). Then:*

(1) *$C_1$ and $C_2$ are subject to (9.17).*
(2) *The solution $P$ of the Stein equation (8.4) that is uniquely specified by its lowest block row*

$$L := (0, \ldots, 0, I_r) P = (\alpha_0, \ldots, \alpha_m), \tag{12.10}$$

*is positive semidefinite.*
(3) *$S$ is a solution of the* $\widehat{\mathbf{aBIP}}(I_n, N, P, C)$.

We remark that condition (1) implies that the Stein equation (8.4) is solvable and hence, by Corollary 10.7, there exists a unique solution $P$ for every specification of the bottom block row.

**Proof of Theorem 12.3:** By assumption, $S$ meets conditions (12.3)–(12.5). Therefore, Theorem 9.1 guarantees that the nontangential limit (9.13) exists, that its value $P_\mathbf{L} = P_S$ is a positive semidefinite solution of the Stein equation (8.4) and that $C_1$ and $C_2$ meet the condition (9.17). Thus, by Corollary 10.7, there is a unique solution $P$ of the Stein equation (8.4) which meets condition (12.10). Moreover, since

$$(P_\mathbf{L})_{mm} \leq \alpha_m = P_{mm} \quad \text{and} \quad (P_\mathbf{L})_{mj} = \alpha_j = P_{mj} \quad (j = 0,\ldots,m-1), \quad (12.11)$$

it follows from Corollary 10.7 that

$$(P_\mathbf{L})_{ij} = P_{ij} \quad \text{for} \quad (i,j) \neq (m,m) \quad \text{and} \quad (P_\mathbf{L})_{mm} \leq P_{mm}. \quad (12.12)$$

Therefore, $0 \leq P_\mathbf{L} \leq P$ and hence, $S \in \widehat{\mathcal{S}}(I_n, N, P, C)$ by Theorem 12.2. □

The first two statements of the last theorem exhibit necessary conditions for the $\widehat{\mathbf{CFBP}}$ to be solvable. The next theorem shows that these conditions are also sufficient. In fact, in view of Theorem 10.5, it is not necessary to assume both of these conditions, since the second automatically implies the first.

THEOREM 12.4. *Let $\xi_j \in \mathbb{C}^{p \times r}$, $\eta_j \in \mathbb{C}^{q \times r}$ and $\alpha_j \in \mathbb{C}^{r \times r}$ be given for $j = 0, \ldots, m$, let $\beta \in \mathbb{T}$, let $N$ be of the form (9.1), let $C_1$ and $C_2$ be partitioned as in (9.2) and suppose that there exists a positive semidefinite solution $P$ of the Stein equation (8.4) with*

$$P_{mj} = \alpha_j, \quad j = 0, \ldots, m.$$

*Then the $\widehat{\mathbf{CFBP}}$ and the $\widehat{\mathbf{aBIP}}(I_n, N, P, C)$ are equivalent (i.e., they have the same set of solutions).*

PROOF. It was shown in the previous theorem that every solution of the $\widehat{\mathbf{CFBP}}$ is a solution of the $\widehat{\mathbf{aBIP}}(I_n, N, P, C)$.

Conversely, if $S \in \widehat{\mathcal{S}}(I_n, N, P, C)$, then, by Theorem 12.2, the nontangential limit (9.13) exists and its value $P_\mathbf{L}$ is subject to the constraints $0 \leq P_\mathbf{L} \leq P$. Moreover, by Theorem 9.1, $P_\mathbf{L}$ satisfies the Stein equation (8.4) and hence, by Corollary 10.8,

$$\angle \lim_{z,\omega \to \beta} \mathbf{L}_{mj}(z,\omega) = (P_\mathbf{L})_{mj} = P_{mj} = \alpha_j$$

for $j = 0, \ldots, m-1$, and

$$\angle \lim_{z,\omega \to \beta} \mathbf{L}_{mm}(z,\omega) = (P_\mathbf{L})_{mm} \leq P_{mm} = \alpha_m.$$

Thus, $S$ satisfies the conditions (12.4) and (12.5) Finally, since Theorem 9.1 also guarantees the existence of the limit (9.15), which is equivalent to (12.3), $S$ is a solution of the $\widehat{\mathbf{CFBP}}$. □

The next lemma helps to clarify the interpolation meaning of the matrices $\alpha_0, \ldots, \alpha_m$.

LEMMA 12.5. *Let $N$ and $C$ be as in (9.1) and (9.2) and let $S \in \mathcal{S}^{p \times q}$ admit the nontangential asymptotic expansion*

$$\sum_{j=0}^{2m+1} (z-\beta)^j \xi_j^* S(z) - \sum_{j=0}^{2m+1} (z-\beta)^j \eta_j^* = o((z-\beta)^{2m+1}), \quad (12.13)$$

in each Stoltz angle $U_\beta(\phi)$ with half angle $\phi < \frac{\pi}{2}$ for some choice of $\xi_{m+1}, \ldots, \xi_{2m+1} \in \mathbb{C}^{p \times r}$ and $\eta_{m+1}, \ldots, \eta_{2m+1} \in \mathbb{C}^{q \times r}$ and assume that (9.16) and (9.17) are in force. Then:

(1) The nontangential limit (9.18) exists and its value $P_\mathbf{V}$ is a solution of the Stein equation (8.4) and is given by the formula

$$P_\mathbf{V} = \mathbf{H}_\eta^* \mathbf{DC}_2 - \mathbf{H}_\xi^* \mathbf{DC}_1, \qquad (12.14)$$

where the matrices $\mathbf{H}_\xi$, $\mathbf{H}_\eta$, $\mathbf{C}_j$ and $\mathbf{D}$ are defined in formulas (10.28), (10.29), (10.7) and (10.1), respectively.

(2) If moreover, $S \in \widehat{\mathcal{S}}(I_n, N, P, C)$, where $P$ is any positive semidefinite solution of the Stein equation (8.4), then the nontangential limit (9.13) exists and its value $P_\mathbf{L}$ is equal to $P_\mathbf{V}$.

PROOF. Upon reexpressing (12.13) in terms of the polynomials (12.1), we see that

$$\frac{\mathcal{B}_\ell(z) - \mathcal{A}_\ell(z) S(z)}{(z - \beta)^{\ell+1}} = \sum_{j=0}^m (z - \beta)^j \xi_{\ell+j+1}^* S(z) - \sum_{j=0}^m (z - \beta)^j \eta_{\ell+j+1}^* + o((z - \beta)^m)$$

for $\ell = 0, \ldots, m$. Therefore, by (9.5),

$$H(z)^{-1}(C_2^* - C_1^* S(z)) = -\begin{pmatrix} \dfrac{\mathcal{A}_0(z) S(z) - \mathcal{B}_0(z)}{(z-\beta)} \\ \vdots \\ \dfrac{\mathcal{A}_m(z) S(z) - \mathcal{B}_m(z)}{(z-\beta)^{m+1}} \end{pmatrix}$$

$$= \sum_{j=0}^m (z-\beta)^j \left( \begin{pmatrix} \xi_{j+1}^* \\ \vdots \\ \xi_{m+j+1}^* \end{pmatrix} S(z) - \begin{pmatrix} \eta_{j+1}^* \\ \vdots \\ \eta_{m+j+1}^* \end{pmatrix} \right) + o((z-\beta)^m).$$

Making use of (10.28), (10.29) and of the mvf

$$X_r(z) = \begin{pmatrix} I_r \\ (z-\beta) I_r \\ \vdots \\ (z-\beta)^m I_r \end{pmatrix},$$

we may rewrite the last asymptotic equality as

$$H(z)^{-1}(C_2^* - C_1^* S(z)) = \mathbf{H}_\xi^* X_p(z) S(z) - \mathbf{H}_\eta^* X_q(z) + o((z-\beta)^m), \qquad (12.15)$$

Next, since

$$\lim_{z \to \beta} \frac{X^{(j)}(z)}{j!} = T^{*j} E^*,$$

Leibnitz's rule and the formulas (9.29), (10.5) lead readily to the evaluation

$$
\begin{aligned}
\angle \lim_{z \to \beta} \frac{1}{m!} \frac{d^m}{dz^m} \left( X_p(z) S(z) C_2 \widetilde{G}(z) \right) &= \angle \lim_{z \to \beta} \sum_{j=0}^{m} \frac{X_p^{(j)}(z)(S(z) C_2 \widetilde{G}(z))^{(m-j)}}{j!(m-j)!} \\
&= \sum_{j=0}^{m} T^{*j} E^*(-1)^{j+1} C_1 (\beta T)^j N^{-j-1} \\
&= \sum_{j=0}^{m} T^{*j} E^*(-1)^{j+1} E(\beta T)^j N^{-j-1} \mathbf{C}_1 \\
&= -\mathbf{D}\mathbf{C}_1, \qquad (12.16)
\end{aligned}
$$

where the passage to the last line uses the fact that $N^{-j-1}$ is a polynomial in $T$. Similarly, on account of (9.9), we have

$$
\begin{aligned}
\angle \lim_{z \to \beta} \frac{1}{m!} \frac{d^m}{dz^m} \left( X_q(z) C_2 \widetilde{G}(z) \right) &= \angle \lim_{z \to \beta} \sum_{j=0}^{m} \frac{1}{j!(m-j)!} X_q^{(j)}(z)(C_2 \widetilde{G}(z))^{(m-j)} \\
&= \sum_{j=0}^{m} T^{*j} E^*(-1)^{j+1} C_2 (\beta T)^j N^{-j-1} \\
&= \sum_{j=0}^{m} T^{*j} E^*(-1)^{j+1} E(\beta T)^j N^{-j-1} \mathbf{C}_2 \\
&= -\mathbf{D}\mathbf{C}_2. \qquad (12.17)
\end{aligned}
$$

Now it follows readily from (12.15)–(12.17) by Lemma 7.8, that

$$
\angle \lim_{z \to \beta} \frac{1}{m!} \frac{d^m}{dz^m} \left( H(z)^{-1}(C_2^* - C_1^* S(z)) C_2 \widetilde{G}(z) \right) = \mathbf{H}_\eta^* \mathbf{D}\mathbf{C}_2 - \mathbf{H}_\xi^* \mathbf{D}\mathbf{C}_1. \qquad (12.18)
$$

Thus, as $C_1$ and $C_2$ are assumed to be subject to (9.17), Theorem 10.5 guarantees that

$$
\frac{d^m}{dz^m} \left( \widetilde{H}(z) \{ C_1^* C_1 - C_2^* C_2 \} G(z)^{-1} \right) = 0
$$

and hence, that

$$
\frac{d^m}{dz^m} \left( \widetilde{H}(z)(C_2^* - C_1^* S(z)) C_2 G(z)^{-1} \right) = \frac{d^m}{dz^m} \left( \widetilde{H}(z) C_1^* (C_1 - S(z) C_2) G(z)^{-1} \right). \qquad (12.19)
$$

Formula (12.14) now follows directly from the definition of $P_\mathbf{V}$ and the formulas (12.18) and (12.19).

Moreover, if $S \in \widehat{\mathcal{S}}(I_n, N, P, C)$, then the limit $P_\mathbf{L}$ in (9.13) exists, by Theorem 12.2. Theorem 9.8 guarantees that $P_\mathbf{L} = P_\mathbf{V}$, which completes the proof of the lemma. □

## 12. A CARATHÉODORY-FEJÉR BOUNDARY PROBLEM

COROLLARY 12.6. *Let $S$ be a solution of the $\widehat{\mathbf{CFBP}}$ that admits a nontangential asymptotic expansion of the form (12.13). Then*

$$(\alpha_0, \ldots, \alpha_{m-1}) = (\eta^*_{m+1}, \ldots, \eta^*_{2m+1}) \mathbf{D} \begin{pmatrix} \eta_0 & \cdots & \eta_{m-1} \\ 0 & \ddots & \vdots \\ \vdots & \ddots & \eta_0 \\ 0 & \cdots & 0 \end{pmatrix}$$

$$- (\xi^*_{m+1}, \ldots, \xi^*_{2m+1}) \mathbf{D} \begin{pmatrix} \xi_0 & \cdots & \xi_{m-1} \\ 0 & \ddots & \vdots \\ \vdots & \ddots & \xi_0 \\ 0 & \cdots & 0 \end{pmatrix}$$

*and*

$$\alpha_m \geq (\eta^*_{m+1}, \ldots, \eta^*_{2m+1}) \mathbf{D} \begin{pmatrix} \eta_m \\ \vdots \\ \eta_0 \end{pmatrix} - (\xi^*_{m+1}, \ldots, \xi^*_{2m+1}) \mathbf{D} \begin{pmatrix} \xi_m \\ \vdots \\ \xi_0 \end{pmatrix}.$$

PROOF. Under the given assumptions, (9.16) and (9.17) are in force and $P_\mathbf{L} = P_\mathbf{V}$, thanks to Theorem 9.1. Thus, the asserted identities follow easily from formulas (12.4), (12.5) and (12.14) upon matching the bottom block rows in $P_\mathbf{L}$ and $P_\mathbf{V}$. □

COROLLARY 12.7. *Let $N$ be as in (9.1), let*

$$C = \begin{pmatrix} C_1 \\ C_2 \end{pmatrix} = \begin{pmatrix} \xi_0 & 0 & \cdots & 0 \\ \eta_0 & \eta_1 & \cdots & \eta_m \end{pmatrix},$$

*let*

$$\xi_0^* S(z) - \sum_{j=0}^{2m+1} (z-\beta)^j \eta_j^* = o((z-\beta)^{2m+1}) \qquad (\text{as } z \to \beta \text{ nontangentially}),$$

*for some choice of $\eta_{m+1}, \ldots, \eta_{2m+1} \in \mathbb{C}^{q \times r}$ and assume that (9.17) is in force. Then the nontangential limit (9.18) exists and its value $P_\mathbf{V}$ is a solution of the Stein equation (8.4) and is given by the formula*

$$P_\mathbf{V} = \mathbf{H}_\eta^* \mathbf{D} C_2, \tag{12.20}$$

*where the matrices $\mathbf{H}_\eta$, $C_2$ and $\mathbf{D}$ are defined in formulas (10.29), (10.7) and (10.1), respectively. Moreover, if $S \in \widehat{\mathcal{S}}(I_n, N, P, C)$, where $P$ is any positive semidefinite solution of the Stein equation (8.4), then the nontangential limit (9.13) exists and its value $P_\mathbf{L}$ is equal to $P_\mathbf{V}$.*

PROOF. The assumptions are the same as in the last lemma except that the condition (9.16) has been dropped. However, in the proof of the last lemma, assumption (9.16) only comes into play in the evaluation of $\mathbf{H}_\xi^* X_p(z) S(z)$. If, as in the present setting, $\xi_j = 0$ for $j = 1, \ldots, 2m+1$, then this assumption is not needed because $\mathbf{H}_\xi^* = 0$. Therefore, the conclusions of the last lemma are in force, with $\mathbf{H}_\xi^* = 0$. □

In keeping with the preceding conventions, we shall let **CFBP** denote the problem of finding all functions $S \in \mathcal{S}^{p \times q}$ which satisfy conditions (12.3), (12.4) and (12.5) with equality.

THEOREM 12.8. *Let $T$, $C_1$ and $C_2$ be the matrices defined in (9.3) and (9.2) and let $P$ be the solution of the Stein equation (8.4), that is uniquely specified by its lowest block row via (12.10). Then the **CFBP** has a solution if and only if $P \geq 0$ and*

$$\operatorname{Ker} PT \cap \operatorname{Ker} C \subseteq \operatorname{Ker} P. \qquad (12.21)$$

PROOF. To apply Theorem 6.1, we shall prove the equivalence of the **CFBP** and the **aBIP**$(I_n, N, P, C)$.

Let $S$ be a solution of the **CFBP**. Then, by Theorem 12.3, (9.17) holds, $P$ is a positive semidefinite solution of the Stein equation (8.4) and $S$ is a solution of the $\widehat{\mathbf{aBIP}}(I_n, N, P, C)$. Therefore, Theorem 9.1 is applicable and hence the nontangential limit $P_\mathbf{L}$ in (9.13) exists and is equal to $P_S$. In particular, since $S$ is a solution of the **CFBP**,

$$(P_S)_{mm} = (P_\mathbf{L})_{mm} = \angle \lim_{z,\omega \to \beta} \mathbf{L}_{mm}(z,\omega) = \alpha_m.$$

By (12.10), $P_{mm} = \alpha_m (= (P_S)_{mm})$. Therefore, since $P \geq P_S \geq 0$ and $P_S$ satisfies the same Stein equation as $P$, Corollary 10.8 guarantees that $P_S = P$. Therefore, $S$ is a solution of the **aBIP**$(I_n, N, P, C)$.

Conversely, if $P$ is a positive semidefinite solution of the Stein equation (8.4), then the **aBIP**$(I_n, N, P, C)$ is defined. If $S \in \mathcal{S}(I_n, N, P, C)$, then $(P_S)_{mm} = P_{mm} = \alpha_m$ and, by another application of Theorem 9.1,

$$\angle \lim_{z,\omega \to \beta} \mathbf{L}_{mm}(z,\omega) = (P_\mathbf{L})_{mm} = (P_S)_{mm} = \alpha_m.$$

Thus, $S$ is a solution of the **CFBP** and, by Theorem 6.1, the **aBIP**$(I_n, N, P, C)$ is solvable if and only if $P$ is a positive semidefinite solution of Stein equation (8.4) and

$$\operatorname{Ker} P(I_n - \beta N) \cap \operatorname{Ker} C = \operatorname{Ker} P \cap \operatorname{Ker} PN \cap \operatorname{Ker} C,$$

since $\beta$ is the only point at which $G(z)$ is not invertible. But the last condition is equivalent to (12.21), since $I_n - \beta N = -\beta T$. □

In conclusion we give an alternative condition to (12.21) for the **CFBP** to be solvable which is in the spirit of Theorem 8.5. The proof is obtained by much the same arguments and will be omitted.

THEOREM 12.9. *Let $P$ be a positive semidefinite solution of the Stein equation (8.4) with $N$ specified by (9.1). Then:*

(1) *If $P$ is minimal in the sense that $P$ does not majorize any positive semidefinite block diagonal matrix $A$ of the form*

$$A = \operatorname{diag}\{A_0, \ldots, A_m\} \quad \text{with} \quad A_j = 0_{r \times r} \; (j = 0, \ldots, m-1) \quad \text{and} \quad A \in \mathbb{C}^{r \times r},$$

*then the **CFBP** is solvable and moreover, it is equivalent to the $\widehat{\mathbf{CFBP}}$.*

(2) *If the **CFBP** is solvable and condition (8.23) holds, then $P$ is minimal in the sense described in the first part.*

There is an analogue of Lemma 8.6 in this setting too:

LEMMA 12.10. *Let $p = q = 1$ and let $P$ be a positive semidefinite solution of the Stein equation (8.4) based on the matrices $N$ and $C$ that are specified in (9.1) and (9.2) with $r = 1$ and let $\nu$ be the integer defined in (8.23). Then:*

(1) *$P$ is invertible if and only if $\nu = 0$.*
(2) *$P$ is singular if and only if $\nu = 1$.*

PROOF. The general strategy of the proof is the same as the proof of Lemma 8.6. Indeed, it is readily seen that the main ingredient is to show that if $\nu = 0$, then $\operatorname{Ker} P = \{0\}$. To verify this claim, let $x$ be a vector in $\operatorname{Ker} P$. Then, since $\nu = 0$ implies that
$$\operatorname{Ker} P = \operatorname{Ker}(P + C_2^* C_2) = \operatorname{Ker}(N^* P N + C_1^* C_1),$$
and since $N = \bar{\beta} I_n + T$, it is readily seen that
$$Px = 0 \implies PTx = 0 \quad \text{and} \quad C_1 x = 0$$
and hence that $PT^j x = 0$ and
$$C_1 T^j x = 0 \quad \text{for} \quad j = 0, 1, \ldots, m.$$
Therefore, since the pair $(C_1, T)$ is observable, it follows that $x = 0$. $\square$

## 13. The full matrix Carathéodory-Fejér boundary problem

In this section we apply the preceding analysis to the Carathéodory-Fejér full matrix boundary problem ($\widehat{\mathbf{CFFP}}$) for $p \times q$ matrix valued Schur functions with $p \geq q$.

The $\widehat{\mathbf{CFFP}}$: *Given a point $\beta \in \mathbb{T}$, a $p \times q$ matrix $\xi_0$ of rank $q$ and $q \times q$ matrices $\eta_0, \ldots, \eta_{2m}$ and $\gamma_{2m+1}$, find all Schur functions $S \in \mathcal{S}^{p \times q}$ such that*

(1) *The matrix valued function*
$$F(z,w) = \frac{\partial^{2m}}{\partial z^m \partial \bar{w}^m} \xi_0^* \left( \frac{I_p - S(z)S(w)^*}{1 - z\bar{w}} \right) \xi_0$$
*meets the uniform bound*
$$\|F(z,z)\| \leq k < \infty \tag{13.1}$$
*in some nontangential neighborhood of $\beta$.*

(2) *The nontangential limit*
$$\eta_{2m+1}^* := \angle \lim_{z \to \beta} (z - \beta)^{-(2m+1)} \left\{ \xi_0^* S(z) - \eta_0^* - \cdots - (z - \beta)^{2m} \eta_{2m}^* \right\} \tag{13.2}$$
*exists and satisfies the bound*
$$(-1)^m \beta^{2m+1} \left( \gamma_{2m+1}^* - \eta_{2m+1}^* \right) \eta_0 \geq 0. \tag{13.3}$$

If (13.2) is in effect, then, by Corollary 7.9, the boundary limits of the first $2m + 1$ derivatives of the function $\xi_0^* S(z)$ exist and satisfy
$$\angle \lim_{z \to \beta} \frac{\xi_0^* S^{(j)}(z)}{j!} = \eta_j^* \quad \text{for} \quad j = 0, \ldots, 2m + 1. \tag{13.4}$$

Our first main objective is to identify the $\widehat{\mathbf{CFFP}}$ with the $\widehat{\mathbf{aBIP}}(I_n, N, P, C)$ based on a matrix $P$ that will be defined below (see Theorem 13.3) and the matrices

$$N = \begin{pmatrix} \bar{\beta}I_q & I_q & & \\ & \bar{\beta}I_q & \ddots & \\ & & \ddots & I_q \\ & & & \bar{\beta}I_q \end{pmatrix} \quad \text{and} \quad C = \begin{pmatrix} C_1 \\ C_2 \end{pmatrix} = \begin{pmatrix} \xi_0 & 0 & \cdots & 0 \\ \eta_0 & \eta_1 & \cdots & \eta_m \end{pmatrix},$$
(13.5)

where $\xi_0$ is a $p \times q$ matrix of rank $q$ and $\eta_j \in \mathbb{C}^{q \times q}$ for $j = 0, \ldots, m$. For this choice of data, the pair $(C, N)$ is clearly observable. A couple of remarks are in order.

REMARK 13.1. A mvf $S$ satisfies (13.2) if and only if it satisfies the nontangential asymptotic equality

$$\xi_0^* S(z) = \eta_0^* + \ldots + (z - \beta)^{2m} \eta_{2m}^* + (z - \beta)^{2m+1} \eta_{2m+1}^* + o\left((z - \beta)^{2m+1}\right). \quad (13.6)$$

Moreover, if (13.6) and (9.17) are in force, then the nontangential limit (9.18) in the setting of (13.5) exists and its value $P_\mathbf{V}$ is given by the formula

$$P_\mathbf{V} = \mathbf{H}_\eta^* \mathbf{D} \mathbf{C}_2, \quad (13.7)$$

where $\mathbf{D}$, $\mathbf{C}_2$ and $\mathbf{H}_\eta$ are the matrices defined in (10.1), (10.7) and (10.29), respectively.

PROOF. The first statement is immediate from Corollary 7.9 applied to the function $\xi_0^* S(z)$; the second follows from Corollary 12.7. □

REMARK 13.2. Let $N$, $C_1$, and $C_2$ be the matrices defined in (13.5) and let relation (9.17) be in force. Then
  (1) The matrices $\eta_0$ and $\mathbf{C}_2$ are invertible.
  (2) The matrix $P_\mathbf{V}$ of the form (13.7) satisfies the Stein equation (8.4) for every choice of the block entries $\eta_{m+1}, \ldots, \eta_{2m+1}$ in $\mathbf{H}_\eta$.

PROOF. By Theorem 10.5, the identity (10.13) is in force. Thus, upon matching the left hand bottom blocks in this identity, we obtain

$$\beta \xi_0^* \xi_0 = \bar{\beta} \eta_0^* \eta_0.$$

Therefore, since rank $\xi_0 = q$ and $\eta_0 \in \mathbb{C}^{q \times q}$, it follows that $\eta_0$ and $\mathbf{C}_2$ are invertible.

The second statement follows from Lemma 10.9 applied to the case $\xi_j = 0$ for $j = 1, \ldots, 2m + 1$. □

THEOREM 13.3. Let

$$P = \mathbf{H}_\gamma \mathbf{D} \mathbf{C}_2, \quad (13.8)$$

where $\mathbf{D}$ and $\mathbf{C}_2$ are the matrices defined in (10.1) and (10.7), respectively, and

$$\mathbf{H}_\gamma = \begin{pmatrix} \gamma_1 & \gamma_2 & \cdots & & \gamma_{m+1} \\ \gamma_2 & \gamma_3 & \cdots & & \vdots \\ \vdots & \vdots & & & \gamma_{2m} \\ \gamma_{m+1} & \gamma_{m+2} & \cdots & \gamma_{2m} & \gamma_{2m+1} \end{pmatrix} \quad \text{with } \gamma_j = \eta_j \text{ for } j = 1, \ldots, 2m. \quad (13.9)$$

Then a mvf $S \in \mathcal{S}^{p \times q}$ is a solution of the $\widehat{\mathbf{CFFP}}$ if and only if the matrix $P$ is a positive semidefinite solution of the Stein equation (8.4) and $S$ is is a solution of the $\widehat{\mathbf{aBIP}}(I_n, N, P, C)$ based on the matrices defined in (13.8) and (13.5).

PROOF. Suppose first that $S$ is a solution of the $\widehat{\mathbf{CFFP}}$ and let $P$, $N$, $C_1$ and $C_2$ be the matrices defined in (13.8) and (13.5). Then, since $\mathcal{A}_m(z) = \xi_0^*$ in the present setting, it is readily seen with the help of formula (9.46) that condition (13.1) coincides with condition (9.14) and hence that all the conclusions of Theorem 9.1 are valid. In particular, $C_1$ and $C_2$ satisfy relation (9.17), the nontangential limits (9.18) for $P_\mathbf{V}$ and (9.13) for $P_\mathbf{L}$ exist and $P_\mathbf{V} = P_\mathbf{L} = P_S$ is a positive semidefinite solution of the Stein equation (8.4). Moreover, by Remark 13.1,

$$P_S = P_\mathbf{L} = P_\mathbf{V} = \mathbf{H}_\eta^* \mathbf{DC}_2. \tag{13.10}$$

Next, comparing formulas (13.8) and (13.10), we see that

$$P - P_S = \begin{pmatrix} 0 & \cdots & 0 & 0 \\ \vdots & & \vdots & \vdots \\ 0 & \cdots & 0 & 0 \\ 0 & \cdots & 0 & \gamma_{2m+1}^* - \eta_{2m+1}^* \end{pmatrix} \mathbf{DC}_2$$

and consequently, as $\mathbf{D}$ and $\mathbf{C}_2$ are both block upper triangular and the lower right hand block entry of $\mathbf{DC}_2$ is equal to $(-1)^m \beta^{2m+1} \eta_0$, that

$$(P - P_S)_{ij} = \begin{cases} 0 & \text{if } (i,j) \neq (m,m), \\ (-1)^m \beta^{2m+1} (\gamma_{2m+1}^* - \eta_{2m+1}^*)\eta_0 & \text{if } (i,j) = (m,m). \end{cases} \tag{13.11}$$

Thus, in view of (13.3), $P_S \leq P$ and hence, by Corollary 10.8, $P$ is also a positive semidefinite solution of the same Stein equation. By Corollary 1.4, $S$ is a solution of the $\widehat{\mathbf{aBIP}}(I_n, N, P, C)$, as claimed.

To prove the opposite implication, assume that the matrix $P$ given in (13.8) is a positive semidefinite solution of the Stein equation (8.4) and that $S$ is a solution of the $\widehat{\mathbf{aBIP}}(I_n, N, P, C)$. Then, since (9.17) and (10.13) hold by Theorem 10.5, $\eta_0$ and $\mathbf{C}_2$ are invertible by Remark 13.2.

Next, from the arguments used to prove Theorem 12.2, it follows that the mvf

$$W(z) = H(z)^{-1} \left\{ C_1^* \left( C_1 - S(z) C_2 \right) G(z)^{-1} - P \right\} \tag{13.12}$$

admits a representation of the form

$$W(z) = \frac{1}{\pi} \int_0^{2\pi} \frac{d\sigma(t)}{e^{it} - z}$$

for some finite positive semidefinite $n \times n$ matrix measure $\sigma(t)$ on $[0, 2\pi)$ and hence, by Lemma 7.2, that

$$\angle \lim_{z \to \beta} (z - \beta) W(z) = -\frac{1}{\pi} \sigma(\{t_0\}) \qquad (\beta = e^{it_0}). \tag{13.13}$$

Upon making use of relations (10.10) and formula (13.8) for $P$, we may rewrite (13.12) as

$$W(z) = H(z)^{-1} \left\{ C_1^* S(z) \mathbf{E}(z) \mathbf{DC}_2 - C_1^* \mathbf{E}(z) \mathbf{DC}_1 - H_\eta \mathbf{DC}_2 \right\}.$$

Moreover, since $C_1 = \xi_0 E$ and $\mathbf{C}_1^* = \mathbf{U}\mathbf{C}_1^*\mathbf{U}$ in the present setting, it is readily seen with the help of (10.13) that

$$\begin{aligned} C_1^* \mathbf{E}(z) \mathbf{D} \mathbf{C}_1 &= E^* \xi_0^* \mathbf{E}(z) \mathbf{D} \mathbf{C}_1 \\ &= E^* \mathbf{E}(z) C_1^* \mathbf{D} \mathbf{C}_1 \\ &= E^* \mathbf{E}(z) \mathbf{U} \mathbf{C}_1^* \mathbf{U} \mathbf{D} \mathbf{C}_1 \\ &= E^* \mathbf{E}(z) \mathbf{U} \mathbf{C}_2^* \mathbf{U} \mathbf{D} \mathbf{C}_2 \end{aligned}$$

and therefore, that

$$W(z) = H(z)^{-1} \left\{ C_1^* S(z) \mathbf{E}(z) - E^* \mathbf{E}(z) \mathbf{U} \mathbf{C}_2^* \mathbf{U} - \mathbf{H}_\gamma \right\} \mathbf{D} \mathbf{C}_2.$$

Since the matrix $\mathbf{D} \mathbf{C}_2$ is invertible, it follows by (13.13), that the nontangential limit

$$\begin{aligned} \Gamma &:= \angle \lim_{z \to \beta} (z - \beta) W(z) (\mathbf{D} \mathbf{C}_2)^{-1} \\ &= \angle \lim_{z \to \beta} (z - \beta) H(z)^{-1} \left\{ C_1^* S(z) \mathbf{E}(z) - E^* \mathbf{E}(z) \mathbf{U} \mathbf{C}_2^* \mathbf{U} - \mathbf{H}_\gamma \right\} \end{aligned} \quad (13.14)$$

exists and is equal to

$$\Gamma = -\frac{1}{\pi} \sigma(\{t_0\}) (\mathbf{D} \mathbf{C}_2)^{-1}.$$

Upon invoking the block decompositions of the entries in (13.14), it is easily seen that the bottom right hand $q \times q$ block of the function $(z - \beta) W(z) (\mathbf{D} \mathbf{C}_2)^{-1}$ is equal to

$$\frac{\xi_0^* S(z) - \eta_0^* - (z - \beta) \eta_1^* - \ldots - (z - \beta)^{2m} \eta_{2m}^* - (z - \beta)^{2m+1} \gamma_{2m+1}^*}{(z - \beta)^{2m+1}}$$

and, by the preceding analysis, that it converges to the bottom right hand $q \times q$ block of the matrix $\Gamma$ as $z$ tends to $\beta$ nontangentially:

$$(\Gamma)_{mm} = \angle \lim_{z \to \beta} (z - \beta)^{-2m-1} \left\{ \xi_0^* S(z) - \sum_{j=0}^{2m} (z - \beta)^j \eta_j^* - (z - \beta)^{2m+1} \gamma_{2m+1}^* \right\}.$$

The latter equality implies that $\angle \lim_{z \to \beta} \dfrac{\xi_0^* S^{(j)}(z)}{j!}$ exists for $j = 0, \ldots, 2m+1$ and that

$$\angle \lim_{z \to \beta} \frac{\xi_0^* S^{(j)}(z)}{j!} = \eta_j^* \quad \text{for} \quad j = 0, \ldots, 2m.$$

It does not yield any information about the value of the last limit for $j = 2m+1$. However, in keeping with the preceding notation, we shall set

$$\eta_{2m+1}^* = \angle \lim_{z \to \beta} \frac{\xi_0^* S^{(2m+1)}(z)}{(2m+1)!}.$$

Then, by Corollary 7.9, the nontangential limit (13.2) exists and hence, by Remark 13.1, the nontangential limit (9.18) in the setting of (13.5) exists and its value $P_{\mathbf{V}}$ is given by the formula (13.7). Since $S$ is a solution of the $\widehat{\mathbf{aBIP}}(I_n, N, P, C)$, the second statement in Lemma 12.5 guarantees that the nontangential limit (9.13) exists and is equal to $P_\mathbf{L} = P_S = P_\mathbf{V} = \mathbf{H}_\eta^* \mathbf{D} \mathbf{C}_2$. Thus, the nontangential limit

$$\angle \lim_{z, \omega \to \beta} \mathbf{L}_{mm}(z, \omega) = (P_\mathbf{L})_{mm}$$

of the right bottom block entry of the kernel $\mathbf{L}_\omega(z)$ defined via (9.10) exists. Therefore, since $(m!)^2 \mathbf{L}_{mm}(z,\omega) = F(z,\omega)$ in the present setting, the uniform bound (13.1) follows.

By Theorem 12.2 we have $P_\mathbf{L} \leq P$, which together with Corollary 10.8 imply that the $q \times q$ blocks
$$(P - P_\mathbf{L})_{ij} = \begin{cases} 0 & \text{if } (i,j) \neq (m,m) \\ \Delta \geq 0 & \text{if } (i,j) = (m,m). \end{cases}$$

Thus, on account of (13.8) and the block upper triangular structure of $\mathbf{DC}_2$,
$$\begin{aligned}\left(\mathbf{H}_\gamma^* - \mathbf{H}_\eta^*\right)_{ij} &= \left((P - P_\mathbf{L})(\mathbf{DC}_2)^{-1}\right)_{ij} \\ &= \begin{cases} 0 & \text{if } (i,j) \neq (m,m), \\ (-1)^m \bar{\beta}^{2m+1} \Delta \eta_0^{-1} & \text{if } (i,j) = (m,m). \end{cases}\end{aligned}$$

In particular,
$$\gamma_{2m+1}^* - \eta_{2m+1}^* = (-1)^m \bar{\beta}^{2m+1} \Delta \eta_0^{-1},$$
which leads easily to (13.3). $\square$

As a simple consequence of Theorem 13.3 we obtain necessary and sufficient conditions for the $\widehat{\mathbf{CFBP}}$ to have a solution.

THEOREM 13.4. *The $\widehat{\mathbf{CFFP}}$ has a solution if and only if the matrix $P$ defined in (13.8) is a positive semidefinite solution of the Stein equation (8.4):*
$$P \geq 0 \quad \text{and} \quad P - N^* P N = C_1 C_1^* - C_2 C_2^*. \tag{13.15}$$
*Moreover, if these conditions are met, then Theorem 5.9 provides a description of the set of all solutions of the $\widehat{\mathbf{CFFP}}$ in terms of a linear fractional transformation with coefficients expressed explicitly in terms of the interpolation data.*

PROOF. The necessity part was proved in Theorem 13.3. Conversely, if $P$ is a positive semidefinite solution of the Stein equation (8.4), then the $\widehat{\mathbf{CFFP}}$ is equivalent to the $\widehat{\mathbf{aBIP}}(I_n, N, P, C)$, which has a solution, by Theorem 5.9. $\square$

EXAMPLE 13.5. The choice $p = q = 1$, $m = 2$, $\beta = 1$, $\xi_0 = \eta_0 = \eta_1 = 1$, $\eta_2 = \eta_3 = \eta_4 = 0$ and $\gamma_5 = \delta > 0$ in formula (13.8) leads to a positive semidefinite singular solution of the Stein equation (8.4):
$$P = \begin{pmatrix} 1 & 0 & 0 \\ 0 & 0 & 0 \\ 0 & 0 & \gamma \end{pmatrix}.$$

For this choice of data, the matrices defined in (5.14) and (5.16) are readily seen to be equal to
$$W_1 = \begin{pmatrix} 1 & 0 & 0 \\ 0 & 0 & 0 \\ 0 & 0 & \delta^{1/2} \\ 1 & 1 & 0 \end{pmatrix}, \ W_2 = \begin{pmatrix} 1 & 1 & 0 \\ 0 & 0 & 0 \\ 0 & 0 & \delta^{1/2} \\ 1 & 0 & 0 \end{pmatrix}, \ X = \begin{pmatrix} 0 \\ 1 \\ 0 \end{pmatrix}, \ Y_1 = Y_2 = \begin{pmatrix} 0 \\ 0 \\ 0 \end{pmatrix}$$
and $Z_1 = Z_2 = 0$. Correspondingly, $\mathbf{Q} = I_3$ and
$$\Delta_0^{[-1]}(z) = \begin{pmatrix} 1 & z-1 & 0 \\ -1 & 2-z & 0 \\ 0 & 0 & \{(1-z)\delta\}^1 \end{pmatrix}.$$

Thus, formula (5.17) for the set of all solutions to this problem reduces to the single solution $S(z) = \Psi_{12}(z) = z$.

We remark that if the $\widehat{\mathbf{CFBP}}$ of Section 12 is specialized to the setting of (13.5), then, since the polynomial $\mathcal{A}_\ell(z) = \xi_0^*$ for $\ell = 0, \ldots, m$, the interpolation conditions (12.3)–(12.5) can be reexpressed as:

$$\angle \lim_{z \to \beta} (z-\beta)^{-m} \left\{ \xi_0^* S(z) - \eta_0^* - (z-\beta)\eta_1^* - \cdots - (z-\beta)^m \eta_m^* \right\} = 0. \quad (13.16)$$

$$\frac{1}{(m!)^2} \angle \lim_{z,\omega \to \beta} \frac{\partial^{2m}}{\partial z^m \partial \bar{\omega}^m} \left( \xi_0^* \Lambda_\omega(z) \xi_0 (\bar{\omega} - \bar{\beta})^{m-\ell} \right) = \alpha_\ell \quad (0 \leq \ell < m). \quad (13.17)$$

$$\frac{1}{(m!)^2} \angle \lim_{z,\omega \to \beta} \frac{\partial^{2m}}{\partial z^m \partial \bar{\omega}^m} \xi_0^* \Lambda_\omega(z) \xi_0 \quad \text{exists and is} \quad \leq \alpha_m. \quad (13.18)$$

The connection between the data $\eta_0, \ldots, \eta_m$ and $\alpha_0, \ldots, \alpha_m$ for the problem $\widehat{\mathbf{CFBP}}$ and the data $\eta_0, \ldots, \eta_{2m}$ and $\gamma_{2m+1}$ for the problem $\widehat{\mathbf{CFFP}}$ is clarified by Corollary 12.6, which, in the setting of (13.5), states that

$$(\alpha_0, \ldots, \alpha_m) = (\eta_{m+1}^*, \ldots, \eta_{2m}^*, \gamma_{2m+1}^*) \mathbf{DC}_2. \quad (13.19)$$

If $p \geq q$ and $\operatorname{rank} \xi_0 = q$, then the $\widehat{\mathbf{CFFP}}$ is equivalent to the problem (13.16)–(13.18), since by the preceding analysis, they are both equivalent to the same $\widehat{\mathbf{aBIP}}(I_n, N, P, C)$ in the setting of (13.5). Moreover, since the matrix $\mathbf{C}_2$ is invertible, the matrices $\eta_{m+1}, \ldots, \eta_{2m}$ and $\gamma_{2m+1}$ are uniquely determined via (13.19) by $\alpha_0, \ldots, \alpha_m$, and vice versa. If $p < q$, one can still formulate the $\widehat{\mathbf{CFFP}}$ via (13.2) and (13.3), but the resulting problem will be not equivalent to the problem (13.16)–(13.18): every solution $S$ of the $\widehat{\mathbf{CFFP}}$ will satisfy the conditions (13.16)–(13.18) but not conversely.

We remark also that there exist solutions $S$ of the $\widehat{\mathbf{CFFP}}$ for which equality holds in (13.3) if and only if the matrix $P$ defined in (13.8) is a positive semidefinite solution of the Stein equation (8.4) and the condition (12.21), based on the matrices $T$ and $C$ defined in (9.3) and (13.5), is in force. Moreover, this problem, which we shall refer to as the $\mathbf{CFFP}$, has one solution if the integer $\nu$ defined by formula (5.5) meets the condition $\nu = \min(p, q)$ and it has infinitely many solutions if $\nu < \min(p, q)$. In particular, if $P$ is positive definite, then the $\mathbf{CFFP}$ has infinitely many solutions. The verification of these assertions is much the same as the proof of Theorems 12.8 and 12.9. The strategy is to identify the $\mathbf{CFFP}$ with the $\mathbf{aBIP}(I_n, N, P, C)$ that is formulated in terms of the matrices $N$, $P$ and $C$ defined in (13.5) and (13.8). Note also that equality in (13.3) means that the limit $\eta_j^*$ of $(j!)^{-1}\xi_0^* S^{(j)}(z)$, as $z$ tends to $\beta$ nontangentially, is specified for $j = 2m+1$ as well as for $j = 0, \ldots, 2m$. In particular, $\eta_{2m+1}^*$ is equal to the preassigned $q \times q$ matrix $\gamma_{2m+1}^*$ for every solution $S$ of the $\mathbf{CFFP}$.

If $p = q$, $\xi_0 = I_q$ and $\eta_j = S_j^*$ for $j = 0, \ldots, 2m$, then the $\widehat{\mathbf{CFFP}}$ reduces to:

The $\widehat{\mathbf{FP}}$: *Given a point $\beta \in \mathbb{T}$ and $q \times q$ matrices $S_0, \ldots, S_{2m}$ and $\gamma$, find all Schur functions $S \in \mathcal{S}^{q \times q}$ such that*

(1) *The matrix valued function*

$$F(z, w) = \frac{\partial^{2m}}{\partial z^m \partial \bar{w}^m} \left( \frac{I_q - S(z)S(w)^*}{1 - z\bar{w}} \right)$$

## 13. THE FULL MATRIX CARATHÉODORY-FEJÉR BOUNDARY PROBLEM

*meets the uniform bound*

$$\|F(z, z)\| \leq k < \infty \tag{13.20}$$

*in some nontangential neighborhood of $\beta$.*
(2) *The nontangential limit*

$$S_{2m+1} := \angle \lim_{z \to \beta} (z - \beta)^{-(2m+1)} \left\{ S(z) - S_0 - \ldots - S_{2m}(z - \beta)^{2m} \right\} \tag{13.21}$$

*exists and satisfies the bound*

$$(-1)^m \beta^{2m+1} \left( \gamma - S_{2m+1} \right) S_0^* \geq 0. \tag{13.22}$$

In this setting, $N$ is the same as in (13.5), whereas

$$C = \begin{pmatrix} C_1 \\ C_2 \end{pmatrix} = \begin{pmatrix} I_q & 0 & \cdots & 0 \\ S_0^* & S_1^* & \cdots & S_m^* \end{pmatrix}. \tag{13.23}$$

By Theorem 13.4, the $\widehat{\mathbf{FP}}$ has a solution if and only if the matrix

$$P = \begin{pmatrix} S_1 & S_2 & \cdots & S_{m+1} \\ S_2 & S_3 & \cdots & \vdots \\ \vdots & \vdots & & S_{2m} \\ S_{m+1} & S_{m+2} & \cdots & S_{2m} & \gamma \end{pmatrix} \mathbf{D} \begin{pmatrix} S_0^* & S_1^* & \cdots & S_m^* \\ 0 & S_0 & \ddots & \vdots \\ \vdots & & \ddots & S_1^* \\ 0 & \cdots & 0 & S_0^* \end{pmatrix} \tag{13.24}$$

satisfies conditions (13.15).

In [**40**] I. Kovalishina considered another multiple boundary interpolation problem for $q \times q$ matrix valued Schur functions, which she termed the **SK** problem:

**SK**: *Given a point $\beta \in \mathbb{T}$ and $q \times q$ matrices $S_0, \ldots, S_{2m}$ and $\gamma$ with $S_0 S_0^* = I_q$, find all Schur functions $S \in \mathcal{S}^{q \times q}$ that meet the conditions (13.21) and (13.22).*

Kovalishina asserted that the **SK** problem is solvable if and only if the matrix $P$ defined by (13.8) is a positive semidefinite solution of the Stein equation (8.4).[5] However, although these conditions are sufficient, they not necessary, as the following extension of Example 9.9 illustrates.

EXAMPLE 13.6. Let $q = 1$, $m = 1$, $\beta = 1$, $S_0 = 1$, $S_1 = \frac{1}{2}$, $S_2 = S_3 = S_4 = \gamma = 0$, $C_1 = (1, 0, 0)$ and $C_2 = (1, \frac{1}{2}, 0)$. Then the function $S(z) = (1 + z)/2$ solves the problem **SK** with this data, since the limit

$$S_5 := \lim_{z \to 1} (z - 1)^{-5} \left\{ S(z) - \sum_{j=0}^{4} (z - 1)^j S_j \right\} = 0$$

exists and satisfies the bound (13.3). However, the matrix

$$P = \mathbf{H}_\gamma \mathbf{D} C_2 = \begin{pmatrix} \frac{1}{2} & 0 & 0 \\ 0 & 0 & 0 \\ 0 & 0 & 0 \end{pmatrix} \begin{pmatrix} 1 & -1 & 1 \\ 0 & -1 & 2 \\ 0 & 0 & 1 \end{pmatrix} \begin{pmatrix} 1 & \frac{1}{2} & 0 \\ 0 & 1 & \frac{1}{2} \\ 0 & 0 & 1 \end{pmatrix} = \begin{pmatrix} \frac{1}{2} & -\frac{1}{4} & \frac{1}{4} \\ 0 & 0 & 0 \\ 0 & 0 & 0 \end{pmatrix}$$

---

[5]Kovalishina considered a functional identity (16) in [**40**] that turns out to be equivalent to (8.4)

is not positive semidefinite and does not satisfy the Stein equation (8.4). In fact, since

$$\mathbf{C}_1^* \mathbf{UDC}_1 = \begin{pmatrix} 0 & 0 & 1 \\ 0 & -1 & 2 \\ 1 & -1 & 1 \end{pmatrix} \neq \begin{pmatrix} 0 & 0 & 1 \\ 0 & -1 & 2 \\ 1 & -1 & \frac{5}{4} \end{pmatrix} = \mathbf{C}_2^* \mathbf{UDC}_2,$$

Theorem 10.5 implies that the Stein equation (8.4) does not have any solutions at all for this choice of data.

In [40], Kovalishina also presented a parametrization of the set of all "solutions" in terms of a linear fractional transformation under the assumption that the associated Pick matrix $P$ given by (13.24) is positive definite. The parametrization is of the form (5.3) with the free parameter $\mathcal{E} \in \mathcal{S}^{q \times q}$. However, since every function $S$ of the form (5.3) satisfies (13.20) as well as the conditions (13.21) and (13.22), the proposed parametrization describes the set of all solutions of the $\widehat{\mathbf{FP}}$ formulated above, rather than the problem **SK**.

To prove the necessity of the conditions (13.15), Kovalishina assumed that

$$S(z)S(1/\bar{z})^* = I_q \tag{13.25}$$

in some neighborhood of $\beta$. Example 13.6 shows that this assumption is restrictive. However, it suggests another modification of the problem **SK**:

**SK'**: *Given a point $\beta \in \mathbb{T}$ and $q \times q$ matrices $S_0, \ldots, S_{2m}$ and $\gamma$, find all Schur functions $S \in \mathcal{S}^{q \times q}$ that meet conditions (13.21), (13.22) and (13.25).*

The condition (13.20) sits between the conditions (1) of $S_0$ being unitary and (2) of $S(z)$ being unitary on an open arc containing the point $\beta$: (2)$\Longrightarrow$ (13.20) $\Longrightarrow$ (1). A bitangential version of the **SK'** problem for rational generalized Schur functions was considered in [13].

It turns out that conditions (13.15) for $P$ defined by (13.24) are necessary and sufficient for the problem **SK'** to have a solution. The necessity was proved in [40]. Furthermore, if $P$ is positive definite, then it is easily shown that formula (5.3) parametrizes the set of all solutions of the problem **SK'**, as $\mathcal{E}$ varies over the class of $\mathcal{S}^{q \times q}$ functions satisfying

$$\mathcal{E}(z)\mathcal{E}(1/\bar{z})^* = I_q$$

in some neighborhood of $\beta$. If $P$ is positive semidefinite and satisfies (8.4), then, using approximation arguments, one can easily show that the problem **SK'** still has a solution.

## 14. The lossless inverse scattering problem

Given $S \in \mathcal{S}^{p \times q}$, the lossless inverse scattering problem (**LISP**) is to find all $J$–inner mvf's $\Theta$ which are analytic in $\mathbb{D}$ such that

$$(I_p, -S(z))\Theta(z)J\Theta(z)^* \begin{pmatrix} I_p \\ -S(z)^* \end{pmatrix} \geq 0 \quad (|z| < 1). \tag{14.1}$$

It is known that this inequality holds if and only if

$$S(z) = \mathbf{T}_\Theta(\mathcal{E}) := (\theta_{11}(z)\mathcal{E}(z) + \theta_{12}(z))(\theta_{21}(z)\mathcal{E}(z) + \theta_{22}(z))^{-1}$$

for some choice of $\mathcal{E} \in \mathcal{S}^{p \times q}$. This representation exhibits $S$ as the input scattering matrix of a lossless network with chain scattering matrix $\Theta$ and load scattering matrix $\mathcal{E}$, which motivates the term **LISP**.

An explicit construction of all rational solutions of the **LISP** which are analytic in $\overline{\mathbb{D}}$ was given in [**23**, Section 8]. For additional information, references and other directions, see [**3**] and [**2**]. In [**23**, Theorem 8.5] all elementary solutions of the **LISP** with a simple pole at the boundary were constructed. In this section we shall describe all the rational solutions of the **LISP** with an arbitrary number of poles on the boundary, simple or not.

It is readily seen from (14.1) that for every solution $\Theta$ of the **LISP** and every constant $J$–unitary matrix $U$, the function $\Theta(z)U$ is also a solution of the same **LISP**. In particular, we may take $U = \Theta^{-1}(\mu)$ (where $\mu$ is an arbitrary point on $\mathbb{T}$ at which $\Theta$ is analytic) and conclude that every solution $\Theta$ of the **LISP** can be normalized by the condition (4.10). By Theorem 4.5, a $J$–inner rational mvf $\Theta$ admits a realization (4.9) under the normalization condition (4.10). By assumption, $\Theta$ has no poles off the unit circle; therefore, $G(z)$ is invertible off the unit circle and we may assume without loss of generality (see Lemma 2.11 and Remark 2.12) that $M$ is the identity matrix and

$$N = N_1 \oplus \cdots \oplus N_k, \quad \text{where} \quad N_j = \begin{pmatrix} \bar{\beta}_j & 1 & & \\ & \bar{\beta}_j & \ddots & \\ & & \ddots & 1 \\ & & & \bar{\beta}_j \end{pmatrix} \in \mathbb{C}^{(m_j+1) \times (m_j+1)}, \tag{14.2}$$

where $\beta_j \in \mathbb{T}$ and $m_1 + \ldots + m_k + k = n$. Here, and in what follows, we let $Z = Z_1 \oplus \cdots \oplus Z_k$ denote the block diagonal matrix with matrices $Z_1, \ldots, Z_k$ on the main diagonal. Then, in accordance with (1.4) and (1.12),

$$G(z) = I_n - zN = G_1(z) \oplus \cdots \oplus G_k(z) \tag{14.3}$$

and

$$H(z) = zI_n - N^* = H_1(z) \oplus \cdots \oplus H_k(z), \tag{14.4}$$

where

$$G_j(z) = I_{m_j+1} - zN_j \quad \text{and} \quad H_j(z) = zI_{m_j+1} - N_j^*.$$

Following [**23**, Section 8], we shall say that a solution $\Theta$ of the form

$$\Theta(z) = I_{p+q} - \rho_\mu(z) \begin{pmatrix} C_1 \\ C_2 \end{pmatrix} G(z)^{-1} P^{-1} G(\mu)^{-*} (C_1^*, C_2^*) J \tag{14.5}$$

of the **LISP** is an elementary $(C_1, N)$ solution. This means in particular, that

$$\operatorname{span}\{(I_p, 0)f : f \in \mathcal{H}(\Theta)\} = \{C_1(I_n - zN)^{-1}x : x \in \mathbb{C}^n\}. \tag{14.6}$$

There is a converse:

LEMMA 14.1. *Let $\Theta(z)$ be a rational $J$-inner mvf of McMillan degree $n$ and let $(C_1, N)$ be an observable pair. Then (14.6) holds if and only if $\Theta(z)$ can be expressed in the form (14.5) for some $C_2 \in \mathbb{C}^{q \times n}$ and some solution $P > 0$ of the Stein equation (8.4).*

PROOF. Let $\Theta$ be a rational $J$–inner mvf of McMillan degree $n$. By Theorem 4.5, it admits a representation of the form

$$\Theta(z) = I_{p+q} - \rho_\mu(z) \begin{pmatrix} \widetilde{C}_1 \\ \widetilde{C}_2 \end{pmatrix} (\widetilde{M} - z\widetilde{N})^{-1} \widetilde{P}^{-1} (\widetilde{M} - \mu\widetilde{N})^{-*} \left( \widetilde{C}_1^*, \widetilde{C}_2^* \right) J, \quad (14.7)$$

where $\det(\widetilde{M} - z\widetilde{N}) \not\equiv 0$, the columns of $\begin{pmatrix} \widetilde{C}_1 \\ \widetilde{C}_2 \end{pmatrix} (\widetilde{M} - z\widetilde{N})^{-1}$ are linearly independent and $\widetilde{P} > 0$ is a solution of the Stein equation

$$\widetilde{M}^* \widetilde{P} \widetilde{M} - \widetilde{N}^* \widetilde{P} \widetilde{N} = \widetilde{C}_1^* \widetilde{C}_1 - \widetilde{C}_2^* \widetilde{C}_2. \quad (14.8)$$

Consequently,

$$\mathcal{H}(\Theta) = \left\{ \begin{pmatrix} \widetilde{C}_1 \\ \widetilde{C}_2 \end{pmatrix} (\widetilde{M} - z\widetilde{N})^{-1} x : \ x \in \mathbb{C}^n \right\},$$

and the equality (14.6) implies that

$$\left\{ \widetilde{C}_1 (\widetilde{M} - z\widetilde{N})^{-1} x : \ x \in \mathbb{C}^n \right\} = \left\{ C_1 (I_n - zN)^{-1} x : \ x \in \mathbb{C}^n \right\}.$$

Therefore, there exists an $n \times n$ invertible matrix $V$ such that

$$\widetilde{C}_1 (\widetilde{M} - z\widetilde{N})^{-1} = C_1 (I_n - zN)^{-1} V. \quad (14.9)$$

This implies in particular that the mvf $\widetilde{C}_1(\widetilde{M} - z\widetilde{N})^{-1}$ is analytic at zero and hence that $\widetilde{M}$ is invertible (see e.g., [**25**, Section 3]). Thus, without loss of generality, we can in fact assume that $\widetilde{M} = I_n$ and hence (upon setting $z = 0$ in (14.9)) that

$$\widetilde{C}_1 = C_1 V. \quad (14.10)$$

Then, by applying the backward shift operator $R_0$ to the identity (14.9), we obtain

$$C_1 (I_n - zN)^{-1} NV = \widetilde{C}_1 (I_n - z\widetilde{N})^{-1} \widetilde{N} = C_1 (I_n - zN)^{-1} V \widetilde{N}.$$

Therefore, since $(C_1, N)$ is an observable pair,

$$NV = V\widetilde{N}. \quad (14.11)$$

The rest is straightforward: setting

$$C_2 = \widetilde{C}_2 V^{-1} \quad \text{and} \quad P = V^{-*} \widetilde{P} V^{-1}$$

and taking (14.10) and (14.11) into account, we conclude that the formulas (14.7) and (14.5) are equivalent. Moreover, upon multiplying (14.8) by $V^{-1}$ on the right and by $V^{-*}$ on the left, we conclude that $P$ satisfies the Stein equation (8.4). □

To establish necessary and sufficient conditions for the **LISP** to have a solution, we first need to extend Theorem 9.1 to the case of several boundary points.

Let $N$ be decomposed as in (14.2), let

$$\widetilde{G}_j(z) = (z - \beta_j)^{m_j+1} G_j^{-1}(z) \quad \text{and} \quad \widetilde{H}_j(z) = (z - \beta_j)^{m_j+1} H_j^{-1}(z) \quad (14.12)$$

for $j = 1, \ldots, k$ and let

$$\widetilde{G}(z) = \widetilde{G}_1(z) \oplus \cdots \oplus \widetilde{G}_k(z) \quad \text{and} \quad \widetilde{H}(z) = \widetilde{H}_1(z) \oplus \cdots \oplus \widetilde{H}_k(z). \quad (14.13)$$

Furthermore, let $C$ be decomposed conformally with (14.2) as

$$C = \begin{pmatrix} C_1 \\ C_2 \end{pmatrix} = \begin{pmatrix} C_{11} & \cdots & C_{1k} \\ C_{21} & \cdots & C_{2k} \end{pmatrix}, \qquad (14.14)$$

where

$$\begin{pmatrix} C_{1j} \\ C_{2j} \end{pmatrix} = \begin{pmatrix} \xi_{j0} & \cdots & \xi_{j,m_j} \\ \eta_{j0} & \cdots & \eta_{j,m_j} \end{pmatrix}. \qquad (14.15)$$

Next, we introduce the kernel $\mathbf{L}(z,\omega)$ with $j\ell$-th block entry

$$\mathbf{L}_{j\ell}(z,\omega) = \frac{1}{m_j! m_\ell!} \frac{\partial^{m_j+m_\ell}}{\partial z^{m_j} \partial \bar{\omega}^{m_\ell}} \left( \widetilde{H}_j(z) C_{1j}^* \Lambda_\omega(z) C_{1\ell} \widetilde{H}_\ell(\omega)^* \right) \quad (j,\ell = 1,\ldots,k) \qquad (14.16)$$

and the mvf $\mathbf{V}(z)$ with $j\ell$-th block entry

$$\mathbf{V}_{j\ell}(z) = \frac{1}{m_j!} \frac{d^{m_j}}{dz^{m_j}} \left( \widetilde{H}_j(z) C_{1j}^* \left( C_{1\ell} - S(z) C_{2\ell} \right) G_\ell^{-1}(z) \right). \qquad (14.17)$$

The kernel $\mathbf{L}(z,\omega)$ is positive on $\mathbb{D} \times \mathbb{D}$, by Proposition 2.4, and the bottom right hand corner $(\mathbf{L}_{jj})_{m_j m_j}(z,\omega)$ of $\mathbf{L}_{jj}(z,\omega)$ is equal to

$$(\mathbf{L}_{jj})_{m_j m_j}(z,\omega) = \frac{1}{(m_j!)^2} \frac{\partial^{2m_j}}{\partial z^{m_j} \partial \bar{\omega}^{m_j}} \left( \mathfrak{a}_j(z) C_{1j}^* \Lambda_\omega(z) C_{1j} \mathfrak{a}_j(\omega)^* \right) \quad (j = 1,\ldots,k),$$

where the

$$\mathfrak{a}_j(z) = \sum_{i=0}^{m_j} \xi_{ji}^* (z - \beta_j)^i, \quad j = 1,\ldots,k, \qquad (14.18)$$

are $1 \times p$ vector polynomials of degree $m_j$ that correspond to the Jordan cells in the decomposition (14.2).

THEOREM 14.2. *Let* $S \in \mathcal{S}^{p \times q}$, *let* $C_{1j} \in \mathbb{C}^{p \times (m_j+1)}$, *let* $\widetilde{H}_j(z)$, $\widetilde{G}_j(z)$ *and* $\mathbf{L}(z,\omega)$ *be defined via* (14.12) *and* (14.16), *respectively, and suppose that*

$$\left\| (\mathbf{L}_{jj})_{m_j m_j}(z,z) \right\| < \kappa < \infty \qquad (14.19)$$

for every point $z$ in a nontangential neighborhood $\mathcal{U}_{\beta_j}$ of $\beta_j$, for $j = 1, \ldots, k$. Then:

(1) $\angle \lim_{z \to \beta_j} \dfrac{1}{m_j!} \dfrac{d^{m_j}}{dz^{m_j}} \left( \widetilde{H}_j(z) C_{1j}^* S(z) \right) = C_{2j}^*$ $(j = 1, \ldots, k)$. (14.20)

(2) $\angle \lim_{z \to \beta_j} \dfrac{1}{m_j!} \dfrac{d^{m_j}}{dz^{m_j}} \left( S(z) C_{2j} \widetilde{G}_j(z) \right) = -C_{1j} N_j^{-1}$ $(j = 1, \ldots, k)$. (14.21)

(3) The nontangential limits
$$[P_\mathbf{V}]_{j\ell} := \angle \lim_{z \to \beta_j} \mathbf{V}_{j\ell}(z) \tag{14.22}$$
and
$$[P_\mathbf{L}]_{j\ell} := \angle \lim_{\substack{z \to \beta_j \\ \omega \to \beta_\ell}} \mathbf{L}_{j\ell}(z, \omega) \tag{14.23}$$
exist for $j, \ell = 1, \ldots, k$ and are equal. Moreover,
$$P_\mathbf{V} = P_\mathbf{L} = P_S, \tag{14.24}$$
where $P_\mathbf{V} = ([P_\mathbf{V}]_{j\ell})_{j,\ell=1}^k$, $P_\mathbf{L} = ([P_\mathbf{L}]_{j\ell})_{j,\ell=1}^k$ and $P_S$ is the matrix associated with $S$ via (1.10).

(4) The columns of the $p \times n$ mvf $B(z) = (C_1 - S(z)C_2)G(z)^{-1}$ (based on the matrices $C_1$ and $C_2$ that are defined in (14.14) and (14.15)) belong to $\mathcal{H}(S)$. Moreover,
$$\langle Bu, Bv \rangle_{\mathcal{H}(S)} = v^* P_S u \tag{14.25}$$
for every choice of vectors $u$ and $v$ in $\mathbb{C}^n$. In particular, $B \in \mathbf{H}_2^{p \times n}$.

(5) The function $\widetilde{B}(\zeta) = H(\zeta)^{-1} \left( C_2^* - C_1^* S(\zeta) \right)$ belongs to $\mathbf{H}_2^{n \times q}$.

PROOF. By Theorem 9.1, the uniform estimates (14.19) imply that the norms of the diagonal blocks of the kernel $\mathbf{L}(z, \omega)$ defined via (14.16) are subject to the uniform bound
$$\|\mathbf{L}_{jj}(z, \omega)\| < \kappa < \infty \quad \text{in } \mathcal{U}_{\beta_j} \times \mathcal{U}_{\beta_j} \quad (j = 1, \ldots, k)$$
and that (14.20) and (14.21) hold; the limits (14.20) serve to define the matrix $C_2 = (C_{21}, \ldots, C_{k1})$.

Theorem 9.1 also guarantees that the mvf's
$$\widetilde{B}_j(\zeta) := H_j(\zeta)^{-1} \left( C_{2j}^* - C_{1j}^* S(\zeta) \right) \quad (j = 1, \ldots, k) \tag{14.26}$$
belong to $\mathbf{H}_2^{(m_j+1) \times q}$, which yields (5), since
$$\widetilde{B}(\zeta) = \mathbf{col}\left( \widetilde{B}_1(\zeta), \ldots, \widetilde{B}_k(\zeta) \right).$$
Next, the columns of the mvf's
$$B_{z,j}(\zeta) = \dfrac{1}{m_j!} \dfrac{\partial^{m_j}}{\partial \bar{z}^{m_j}} \left( \Lambda_z(\zeta) C_{1j} \widetilde{H}_j(z)^* \right) \quad (j = 1, \ldots, k) \tag{14.27}$$
belong to $\mathcal{H}(S)$, since $\Lambda_z(\zeta)$ is the reproducing kernel of $\mathcal{H}(S)$, and
$$\angle \lim_{z \to \beta_j} B_{z,j}(\zeta) = (I_p, -S(\zeta)) C_j G_j(\zeta)^{-1}, \tag{14.28}$$

by Lemma 9.6. Moreover, by assertion (e) of Theorem 9.1, the columns of the mvf's

$$B_j(\zeta) := (I_p, -S(\zeta)) \begin{pmatrix} C_{1j} \\ C_{2j} \end{pmatrix} G_j(\zeta)^{-1} \tag{14.29}$$

belong to $\mathcal{H}(S)$, which yields the first half of (4), since

$$B(\zeta) = (B_1(\zeta), \ldots, B_k(\zeta)). \tag{14.30}$$

The proof of (3) and the rest of (4) is broken into steps that are modeled on the proof of Theorem 9.1.

**Step 1.** *The nontangential limits (14.22) exist and the matrix $P_\mathbf{V}$ defined in (14.24) is positive semidefinite.*

**Proof of Step 1:** First we note that the existence of the limits (14.22) for $j = \ell$ is immediate from Theorem 9.1. Furthermore, in view of (9.49) the bounds

$$\|B_{z,j} x\|_{\mathcal{H}(S)}^2 < \kappa_j \|x\|^2 \qquad (\kappa_j < \infty).$$

hold in a nontangential neighborhood $\mathcal{U}_{\beta_j}$ of $\beta_j$ for $j = 1, \ldots, k$. Therefore, $B_{z,j}$ tends weakly to $B_j(\zeta)$ in $\mathcal{H}(S)$ as $z \to \beta_j$ and hence

$$\begin{aligned}
\langle B_\ell y, B_j x \rangle_{\mathcal{H}(S)} &= \angle \lim_{z \to \beta_j} \langle B_\ell y, B_{z,j} x \rangle_{\mathcal{H}(S)} \\
&= \angle \lim_{z \to \beta_j} \frac{1}{m_j!} \langle B_\ell y, \frac{\partial^{m_j}}{\partial \bar{z}^{m_j}} \left( \Lambda_z C_{1j} \widetilde{H}_j(z)^* \right) x \rangle_{\mathcal{H}(S)} \\
&= \angle \lim_{z \to \beta_j} \frac{1}{m_j!} \frac{d^{m_j}}{dz^{m_j}} \left( x^* \widetilde{H}_j(z) C_{1j}^* B_\ell(z) y \right) \\
&= \angle \lim_{z \to \beta_j} \frac{1}{m_j!} \frac{d^{m_j}}{dz^{m_j}} \left( x^* \widetilde{H}_j(z) C_{1j}^* (C_{1\ell} - S(z) C_{2\ell}) G_\ell^{-1}(z) y \right) \\
&= \angle \lim_{z \to \beta_j} x^* \mathbf{V}_{j\ell}(z) y \tag{14.31}
\end{aligned}$$

for every choice of vectors $x \in \mathbb{C}^{m_j}$ and $y \in \mathbb{C}^{m_\ell}$. This implies that the limits (14.22) all exist and that the full matrix $P_\mathbf{V}$ defined in (14.24) is positive semidefinite.

**Step 2.** *The integral (1.11) converges to a matrix $P_S$ which is equal to $P_\mathbf{L}$.*

**Proof of Step 2:** Since the kernel $\mathbf{L}(z, \omega)$ is positive, its off diagonal blocks $\mathbf{L}_{j\ell}(z, \omega)$ also are uniformly bounded in norm in $\mathcal{U}_{\beta_j} \times \mathcal{U}_{\beta_\ell}$. Thus, there exist $k$ sequences $\{\alpha_{1i}\}, \ldots, \{\alpha_{ki}\}$ of points $\alpha_{ji} \in \mathcal{U}_{\beta_j}$ tending to $\beta_j$ such that the limits

$$\angle \lim_{\substack{\alpha_{ji} \to \beta_j \\ \alpha_{\ell i} \to \beta_\ell}} \mathbf{L}_{j\ell}(\alpha_{ji}, \alpha_{\ell i}) =: (P_\mathbf{L})_{j\ell} \in \mathbb{C}^{m_j \times m_\ell} \quad (j, \ell = 1, \ldots, k) \tag{14.32}$$

all exist. Setting

$$\widehat{z} = (z_1, \ldots, z_k) \in \mathbb{D}^k, \quad \widehat{\omega} = (\omega_1, \ldots, \omega_k) \in \mathbb{D}^k, \quad \widehat{\beta} = (\beta_1, \ldots, \beta_k) \in \mathbb{T}^k$$

and

$$\widehat{\alpha}_i = (\alpha_{i1}, \ldots, \alpha_{ik}) \in \mathcal{U}_{\widehat{\beta}} := \mathcal{U}_{\beta_1} \times \cdots \times \mathcal{U}_{\beta_k},$$

we introduce the matrix

$$\mathbf{L}_{\widehat{z}, \widehat{\omega}} = (\mathbf{L}_{j\ell}(z_j, \omega_\ell))_{j,\ell=1}^k. \tag{14.33}$$

Then the equalities (14.32) can be written in the form

$$P_\mathbf{L} = \lim_{\widehat{\alpha}_i \to \widehat{\beta}} \mathbf{L}_{\widehat{\alpha}_i, \widehat{\alpha}_i}. \tag{14.34}$$

Let us consider the mvf
$$\Psi_{\widehat{z}}(\zeta) = (\Psi_{z_1,1}(\zeta), \ldots, \Psi_{z_k,k}(\zeta)), \tag{14.35}$$
where
$$\Psi_{z,j}(\zeta) = \frac{1}{m_j!}\frac{\partial^{m_j}}{\partial \bar{z}^{m_j}}\left(\begin{pmatrix} I_p \\ S(z)^* \end{pmatrix}\frac{C_{1j}\widetilde{H}_j(z)^*}{\rho_z(\zeta)}\right). \tag{14.36}$$
Since the limits (14.20) exist, Lemma 9.5 guarantees that
$$\angle\lim_{z\to\beta_j}\Psi_{z,j}(\zeta) = \begin{pmatrix} C_{1j} \\ C_{2j} \end{pmatrix}G_j(\zeta)^{-1}\quad\text{for}\quad j=1,\ldots,k \tag{14.37}$$
and therefore, in view of the decompositions (14.3), (14.14) and (14.35), that
$$\angle\lim_{\widehat{z}\to\widehat{\beta}}\Psi_{\widehat{z}}(\zeta) = \begin{pmatrix} C_1 \\ C_2 \end{pmatrix}G(\zeta)^{-1}.$$
Moreover, in view of Lemma 9.6 and formulas (14.16) and (14.36),
$$\mathbf{L}_{j\ell}(z,\omega) = [\Psi_{\omega,\ell},\,\Psi_{z,j}]_S,$$
and hence,
$$\mathbf{L}_{\widehat{\alpha}_i,\widehat{\alpha}_i} = [\Psi_{\widehat{\alpha}_i},\,\Psi_{\widehat{\alpha}_i}]_S.$$
Furthermore, invoking Fatou's lemma, just as in (9.54), we obtain the bound
$$\begin{aligned}P_S &:= \left[\begin{pmatrix} C_1 \\ C_2 \end{pmatrix}G(\zeta)^{-1},\,\begin{pmatrix} C_1 \\ C_2 \end{pmatrix}G(\zeta)^{-1}\right]_S\\ &= \frac{1}{2\pi}\int_0^{2\pi}\lim_{\widehat{\alpha}_i\to\widehat{\beta}}\Psi_{\widehat{\alpha}_i}(e^{it})^*\begin{pmatrix} I_p & -S(e^{it}) \\ -S(e^{it})^* & I_q \end{pmatrix}\Psi_{\widehat{\alpha}_i}(e^{it})dt\\ &\leq \lim_{\widehat{\alpha}_i\to\widehat{\beta}}\frac{1}{2\pi}\int_0^{2\pi}\Psi_{\widehat{\alpha}_i}(e^{it})^*\begin{pmatrix} I_p & -S(e^{it}) \\ -S(e^{it})^* & I_q \end{pmatrix}\Psi_{\widehat{\alpha}_i}(e^{it})dt\\ &= \lim_{\widehat{\alpha}_i\to\widehat{\beta}}[\Psi_{\widehat{\alpha}_i},\,\Psi_{\widehat{\alpha}_i}]_S = \lim_{\widehat{\alpha}_i\to\widehat{\beta}}\widehat{\mathbf{L}}_{\widehat{\alpha}_i,\widehat{\alpha}_i} = P_{\mathbf{L}}.\end{aligned} \tag{14.38}$$
Therefore, since
$$(P_{\mathbf{L}})_{jj} = (P_{\mathbf{V}})_{jj} = (P_S)_{jj}\quad\text{for}\quad j=1,\ldots,k, \tag{14.39}$$
by Theorem 9.1, it follows that $P_{\mathbf{L}} = P_S$.

**Step 3.** *The nontangential limit (14.23) exists and (14.24) and (14.25) are in force.*

**Proof of Step 3:** By Step 2, the integral (1.11) converges to a matrix $P_S$ and therefore, $S$ is a solution of the **aBIP**$(I_n, N, P_S, C)$. By Theorem 3.8, the corresponding mvf $\mathbf{W}$ defined by (3.12) belongs to the Carathéodory class $\mathcal{C}^{n\times n}$ and, by Lemma 3.7, the identity (3.13) holds (with $M = I_n$ and $P = P_S$). Comparing the corresponding $(m_j+1)\times(m_\ell+1)$ block entries in (3.13) we get

$$\begin{aligned}C_{1j}^*\Lambda_\omega(z)C_{1\ell} &= \frac{H_j(z)\left(\mathbf{W}_{j\ell}(z) + \mathbf{W}_{\ell j}(\omega)^* - \widetilde{B}_j(z)\widetilde{B}_\ell(\omega)^*\right)H_\ell(\omega)^*}{\rho_\omega(z)}\\ &\quad + C_{1j}^*B_\ell(z) + B_j(\omega)^*C_{1\ell} - (P_S)_{j\ell},\end{aligned} \tag{14.40}$$

where $B_j(z)$ and $\widetilde{B}_j(z)$ are defined in (14.29) and (14.26), respectively, and
$$\mathbf{W}_{j\ell}(z) = -\tfrac{1}{2}H_j(z)^{-1}\left(N_j^* + zI_{m_j+1}\right)(P_S)_{j\ell} + zH_j(z)^{-1}C_{1j}^*B_\ell(z)$$

14. THE LOSSLESS INVERSE SCATTERING PROBLEM

are the $m_j \times m_\ell$ block entries from the block decomposition of the mvf $\mathbf{W}$. Multiplying both sides of (14.40) by $\widetilde{H}_j(z)$ on the left and by $\widetilde{H}_\ell(\omega)^*$ on the right (these matrices are defined in (14.12)) and applying the operator $\dfrac{1}{m_j!m_\ell!}\dfrac{\partial^{m_j+m_\ell}}{\partial z^{m_j}\partial\bar{\omega}^{m_\ell}}$ to both of sides of the resulting identity, we get (just as in (9.59))

$$\mathbf{L}_{j\ell}(z,\omega) = \mathbf{K}_{j\ell}(z,\omega) + \frac{1}{m_j!}\frac{d^{m_j}}{dz^{m_j}}\left(\widetilde{H}_j(z)C_{1j}^*B_\ell(z)\right)$$
$$+ \frac{1}{m_\ell!}\frac{d^{m_\ell}}{d\bar{\omega}^{m_\ell}}\left(B_j(\omega)^*C_{1\ell}\widetilde{H}_\ell(\omega)^*\right) - (P_S)_{j\ell}, \qquad (14.41)$$

where

$$\mathbf{K}_{j\ell}(z,\omega) = \frac{1}{m_j!m_\ell!}\frac{\partial^{m_j+m_\ell}}{\partial z^{m_j}\partial\bar{\omega}^{m_\ell}}((z-\beta_j)^{m_j+1}$$
$$\times \frac{\mathbf{W}_{j\ell}(z) + \mathbf{W}_{\ell j}(\omega)^* - \widetilde{B}_j(z)\widetilde{B}_\ell(\omega)^*}{\rho_\omega(z)}(\bar{\omega}-\bar{\beta}_\ell)^{m_\ell+1}). \qquad (14.42)$$

Therefore, in view of Step 1,

$$(P_\mathbf{L})_{jj} = \angle \lim_{z,\omega\to\beta_j} \mathbf{K}_{jj}(z,\omega) + 2(P_V)_{jj} - (P_S)_{jj} \quad (j=1,\ldots,k)$$

and hence, by (14.39),

$$\angle \lim_{z,\omega\to\beta_j} \mathbf{K}_{jj}(z,\omega) = 0 \quad \text{for} \quad j=1,\ldots,k. \qquad (14.43)$$

By Theorem 3.8, the kernel

$$\frac{P+zW(z)+\bar{\omega}W(\omega)^* - \widetilde{B}(z)\widetilde{B}(\omega)^*}{\rho_\omega(z)} = \frac{W(z)+W(\omega)^* - \widetilde{B}(z)\widetilde{B}(\omega)^*}{\rho_\omega(z)}$$

is positive on $\mathbb{D}\times\mathbb{D}$ and therefore, by Proposition 2.4, the kernel

$$\mathbf{K}(z,\omega) = (\mathbf{K}_{j\ell}(z,\omega))_{j,\ell=1}^k$$

defined in (14.42) is positive on $\mathbb{D}\times\mathbb{D}$. Thus, (14.43) forces

$$\angle \lim_{\substack{z\to\beta_j \\ \omega\to\beta_\ell}} \mathbf{K}_{j\ell}(z,\omega) = 0 \quad \text{for} \quad j=1,\ldots,k. \qquad (14.44)$$

Consequently,

$$(P_\mathbf{L})_{j\ell} = \angle \lim_{\substack{z\to\beta_j \\ \omega\to\beta_\ell}} \mathbf{L}_{j\ell}(z,\omega) = 2(P_V)_{j\ell} - (P_S)_{j\ell} \quad (j,\ell=1,\ldots,k),$$

which, with the help of Step 2, can be written in matrix form as

$$P_\mathbf{L} = 2P_V - P_S = 2P_V - P_\mathbf{L}. \qquad (14.45)$$

Therefore, $P_\mathbf{L} = P_V = P_S$, which completes the proof of (3) and, with the help of formula (14.31), also completes the proof of (4). □

The last theorem enables us to complete the proof of Lemma 3.5.

**Proof of Lemma 3.5:** In order to complete the proof, it suffices to show that

$$\|Bx\|_{\mathcal{H}(S)}^2 = x^*P_S x$$

for every $x \in \mathbb{C}^n$ when $S$ is a solution of the $\widehat{\mathbf{aBIP}}(I_n, N, P, C)$ and $\operatorname{spec}(N) \subset \mathbb{T}$. But in this setting, condition (14.19) holds by Theorem 12.2. Thus, Theorem 14.2 is applicable and yields the desired equality via (14.25). □

The next theorem establishes necessary and sufficient conditions for $S$ to admit an elementary $(C_1, N)$ solution to the **LISP**.

THEOREM 14.3. *Let $S \in \mathcal{S}^{p \times q}$, let $C_1 \in \mathbb{C}^{p \times n}$ and $N \in \mathbb{C}^{n \times n}$ be the matrices specified in (14.2) and (14.14), respectively. Assume further that $(C_1, N)$ is an observable pair, let $C_2 \in \mathbb{C}^{q \times n}$ and let $P > 0$ be a solution of the Stein equation (8.4). Then the mvf $\Theta$ given by formulas (14.5) and (14.3) is an elementary $(C_1, N)$ solution of the **LISP** if and only if:*

(1) *The limits $P_\mathbf{L}$ and $C_{2j}$ $(j = 1, \ldots, k)$ in (14.23) and (14.20) exist.*
(2) *The parameters $C_2$ and $P$ in the definition of $\Theta$ meet the conditions*

$$C_2 = (C_{21}, \ldots, C_{2k}) \quad \text{and} \quad P \geq P_\mathbf{L}.$$

PROOF. Let $\Theta$ be a solution of the **LISP** of the form (14.5). Then, since $S = T_\Theta(\mathcal{E})$ for some $\mathcal{E} \in \mathcal{S}^{p \times q}$ it is a solution of the $\widehat{\mathbf{aBIP}}(I_n, N, P, C)$, by Theorem 5.3. Thus, in view of the block decompositions (14.2) and (14.14), $S$ is also a solution of the $\widehat{\mathbf{aBIP}}(I_{m_j+1}, N_j, P_{jj}, D_j)$ for $j = 1, \ldots, k$, where $D_j = \begin{pmatrix} C_{1j} \\ C_{2j} \end{pmatrix}$. Replacing $N$, $C$ and $P$ in Theorem 12.2 by $N_j$, $D_j$ and $P_j$, respectively, we conclude that the nontangential limits (14.20) exist and serve to uniquely define the matrix $C_2 = (C_{21}, \ldots, C_{2k})$. Moreover, by the same theorem, the nontangential limits

$$(P_\mathbf{L})_{jj} = \angle\lim_{z,\omega \to \beta_j} \frac{1}{(m_j!)^2} \frac{\partial^{2m_j}}{\partial z^{m_j} \partial \overline{\omega}^{m_j}} \left( \widetilde{H}_j(z) C_{1j}^* \Lambda_\omega(z) C_{1j} \widetilde{H}_j(\omega)^* \right)$$

exist for $j = 1, \ldots, k$. But now as the lowest right hand entry of the matrix

$$\widetilde{H}_j(z) C_{1j}^* \Lambda_\omega(z) C_{1j} \widetilde{H}_j(\omega)^*$$

is equal to $\mathfrak{a}_j(z) \Lambda_\omega(z) \mathfrak{a}_j(\omega)^*$, by definitions (14.12), (9.5) and (14.18), this guarantees that the nontangential limits

$$\angle\lim_{z,\omega \to \beta_j} \frac{1}{(m_j!)^2} \frac{\partial^{2m_j}}{\partial z^{m_j} \partial \overline{\omega}^{m_j}} \left( \mathfrak{a}_j(z) \Lambda_\omega(z) \mathfrak{a}_j(\omega)^* \right)$$

exist for $j = 1, \ldots, k$ and hence also that the uniform bounds (14.19) are in force. Thus, all the statements of Theorem 14.2 hold and, in particular, the limits (14.23) exist for $j, \ell = 1, \ldots, k$ and define the matrix $P_\mathbf{L}$.

Finally, $P_S \leq P$, since $S$ is a solution of the $\widehat{\mathbf{aBIP}}(I_n, N, P, C)$ and $P_\mathbf{L} = P_S$, by another application of Theorem 14.2. Thus, $P_\mathbf{L} \leq P$.

Conversely let conditions (1) and (2) be in force. Then, just as above, we conclude from the existence of $P_\mathbf{L}$ that the uniform bounds (14.19) hold and therefore, by Theorem 14.2, that the integral (1.11) converges to a matrix $P_S$, which is equal to $P_\mathbf{L}$. By the second condition, $P_S = P_\mathbf{L} \leq P$, and therefore, $S$ is a solution of the $\widehat{\mathbf{aBIP}}(I_n, N, P, C)$. By Theorem 5.3, $S = T_\Theta(\mathcal{E})$ for some $\mathcal{E} \in \mathcal{S}^{p \times q}$. Consequently, $\Theta$ is a solution of the given **LISP**. □

With the help of Theorem 14.2 it is readily checked that the limit $P_{\mathbf{L}}$ exists if and only if the uniform bound (14.19) holds and hence (by another application of Theorem 14.2) that the existence of the limit $P_{\mathbf{L}}$ guarantees the existence of the limits $C_{2j}$. Notice that although the block $C_2$ in the definition of $\Theta$ is uniquely defined by $C_1$, $N$ and $S$ for solutions $\Theta$ of the **LISP**, there may be many solutions $P > 0$ of the corresponding Stein equation (8.4). On the other hand, the mere existence of the limit $P_{\mathbf{L}}$ (and hence also the limits $C_{2j}$) does not insure the existence of a solution $\Theta$ of McMillan degree $n$ to the **LISP** (even though $P_{\mathbf{L}}$ is a positive semidefinite solution of the Stein equation (8.4)) because there may not be any solutions $P > 0$ of (8.4).

# Bibliography

1. D. Alpay and V. Bolotnikov, *On tangential interpolation in reproducing kernel Hilbert space modules and applications*, in: Topics in Interpolation Theory (H. Dym, B. Fritzsche, V. Katsnelson and B. Kirstein, eds.), Oper. Theory Adv. Appl., **OT95**, Birkhäuser Verlag, Basel, 1997, pp. 37–68.
2. D. Alpay, P. Dewilde and H. Dym, *On the existence and construction of solutions to the partial lossless inverse scattering problem with applications to estimation theory*, IEEE Trans. Inform. Theory, **35** (1989), 1184–1205.
3. D. Alpay and H. Dym, *Hilbert spaces of analytic functions, inverse scattering and operator models, I*, Integral Equations Operator Theory, **7** (1984), 589–641.
4. D. Alpay and H. Dym, *On a new class of structured reproducing kernel Hilbert spaces*, J. Funct. Anal., **111** (1993), 1–28.
5. D. Alpay and H. Dym, *On a new class of reproducing kernel spaces and a new generalization of the Iohvidov laws*, Linear Algebra Appl., **178** (1993), 109–183.
6. D. Alpay and H. Dym, *On a new class of realization formulas and their application*, Linear Algebra Appl., **241-243** (1996), 3–84.
7. N. Aronszajn, *Theory of reproducing kernels*, Trans. Amer. Math. Soc., **68** (1950), 337-404.
8. D. Z. Arov, *Darlington realization of matrix-valued functions*, Math. USSR Izvestija, **7** (1973), 1295–1326.
9. D. Z. Arov and L. Z. Grossman, *Scattering matrices in the theory of unitary extensions of isometric operators*, Soviet Math. Dokl., **270** (1983), 17–20.
10. D. Z. Arov and L. Z. Grossman, *Scattering matrices in the theory of unitary extensions of isometric operators*, Math. Nachr., **157** (1992), 105–123.
11. J. A. Ball, *Models for non contractions*, J. Math. Anal. Appl., **52** (1975), 235–254.
12. J. A. Ball, *Interpolation problems of Pick–Nevanlinna and Loewner type for meromorphic matrix functions*, Integral Equations Operator Theory, **6** (1983), 804–840.
13. J. A. Ball, I. Gohberg and L. Rodman, *Interpolation of rational matrix functions*, Birkhäuser Verlag, Basel, 1990.
14. J. A. Ball and J. W. Helton, *Interpolation problems of Pick–Nevanlinna and Loewner types for meromorphic matrix-functions: parametrization of the set of all solutions*, Integral Equations Operator Theory, **9** (1986), 155–203.
15. V. Bolotnikov and H. Dym, *On degenerate interpolation maximum entropy and extremal problems for matrix Schur functions*, Integral Equations Operator Theory, **32**, (1998), No. 4, 367–435.
16. L. de Branges, *Some Hilbert spaces of analytic functions I*, Trans. Amer. Math. Soc., **106** (1963), 445–468.
17. L. de Branges and J. Rovnyak, *Canonical models in quantum scattering theory*, in: Perturbation Theory and its Applications in Quantum Mechanics (C. Wilcox, ed.), Holt, Rinehart and Winston, New–York, 1966, p. 295–392.
18. L. de Branges and J. Rovnyak, *Square summable power series*, Holt, Rinehart and Winston, New–York, 1966.
19. C. Carathéodory, *Über die Winkelderivierten von beschränkten Funktionen*, Sitzungber. Preuss. Akad. Wiss. (1929), 39–52.
20. P. Dewilde and H. Dym, *Lossless inverse scattering, digital filters, and estimation theory*, IEEE Trans. Inform. Theory, **30** (1984), no. 4, 644–662.
21. W. F. Donoghue, Jr. *Monotone matrix functions and analytic continuation*, Springer–Verlag, New York–Heidelberg, 1974.

22. V. Dubovoj, *Indefinite metric in the interpolation problem of schur for analytic matrix functions* IV, Theor. Funktsii, Func. Anal. i Prilozen., **42** (1984), 46–57. English transl. in: *Topics in Interpolation Theory* (H. Dym, B. Fritzsche, V. Katsnelson and B. Kirstein, eds.), Oper. Theory Adv. Appl., **OT95**, Birkhäuser Verlag, Basel, 1997, pp. 93–104.
23. H. Dym, *J-Contractive Matrix Functions, Reproducing Kernel Spaces and Interpolation*, CBMS Reg. Conf., Ser. in Math. **vol 71**, Amer. Math. Soc., Providence, RI, 1989.
24. H. Dym, *On reproducing kernel spaces, J unitary matrix functions, interpolation and displacement rank*, in: *Topics in Analysis and Operator Theory* (H. Dym, S. Goldberg, M. A. Kaashoek and P. Lancaster, eds.), Oper. Theory Adv. Appl., **OT41**, Birkhäuser Verlag, Basel, 1989, pp. 173–239.
25. H. Dym, *Shifts, realizations and interpolation, redux*, in: *Operator Theory and its Applications* (A. Feintuch and I. Gohberg, eds.), Oper. Theory Adv. Appl., **OT73**, Birkhäuser Verlag, Basel, 1994, pp. 182–243.
26. H. Dym, *More on maximum entropy interpolants and maximum determinant completions of associated pick matrices*, Integral Equations Operator Theory, **24** (1996), 188–229.
27. H. Dym, *A basic interpolation problem*, in: *Holomorphic Spaces* (D. Sarason, S. Axler and J. McCarthy, eds.), Cambridge University Press, 1998, pp. 381–423.
28. F. R. Gantmacher, *The Theory of Matrices, Vol. II*, Chelsea Publishing Company, New–York, 1959.
29. D. R. Georgijević, *Solvability condition for a boundary value interpolation problem of Loewner type*, J. Anal. Math. **74** (1998), 213–234.
30. J. W. Helton, *The distance of a function to $H^\infty$ in the Poincaré metric; electrical power transfer*, J. Funct. Anal. **38** (1980), no. 2, 273–314.
31. A. Hindmarsh. *Pick conditions and analyticity*, Pacific J. Math. **27** (1968), 527–531.
32. G. Julia, *Extension nouvelle d'un lemme de Schwartz*, Acta Math. **42** (1920), 349–355.
33. V. Katsnelson, *Methods of J-Theory in Continuous Interpolation Problems of Analysis*, Private translation by T. Ando, Sapporo, 1982.
34. V. Katsnelson, *Continuous analogues of the Hamburger–Nevanlinna theorem and fundamental matrix inequalities*, Amer. Math. Soc. Transl., **136** (1987), 49–96.
35. V. Katsnelson, *A transformation of Potapov's fundamental matrix inequality*, in: *Topics in Interpolation Theory* (H. Dym, B. Fritzsche, V. Katsnelson and B. Kirstein, eds.), Oper. Theory Adv. Appl., **OT95**, Birkhäuser Verlag, Basel, 1997, pp. 253–281.
36. V. Katsnelson, A. Kheifets and P. Yuditskii, *An abstract interpolation problem and extension theory of isometric operators*, in: *Operators in Spaces of Functions and Problems in Function Theory* (V.A. Marchenko, ed.), **146**, Naukova Dumka, Kiev, 1987, pp. 83–96. English transl. in: *Topics in Interpolation Theory* (H. Dym, B. Fritzsche, V. Katsnelson and B. Kirstein, eds.), Oper. Theory Adv. Appl., **OT95**, Birkhäuser Verlag, Basel, 1997, pp. 283–298.
37. V. Katsnelson and B. Kirstein, *On the theory of matrix–valued functions belonging to the Smirnov class*, in: *Topics in Interpolation Theory* (H. Dym, B. Fritzsche, V. Katsnelson and B. Kirstein, eds.), Oper. Theory Adv. Appl., **OT95**, Birkhäuser Verlag, Basel, 1997, pp. 299–350.
38. A. Kheifets and P. Yuditskii, *An analysis and extension approach of V. P. Potapov's approach to scheme interpolation problems with applications to the generalized bi-tangential Schur–Nevanlinna–Pick problem and J–inner–outer factorization*, in: *Matrix and Operator Valued Functions* (I. Gohberg and L.A. Sakhnovich, eds.), Oper. Theory Adv. Appl., **OT72**, Birkhäuser Verlag, Basel, 1994, pp. 133–161.
39. I. V. Kovalishina, *Carathéodory–Julia theorem for matrix–functions*, Teoriya Funktsii, Funktsional'nyi Analiz i Ikh Prilozheniya, **43** (1985), 70–82. English translation in: Journal of Soviet Mathematics, **48(2)** (1990), 176–186.
40. I. V. Kovalishina, *A multiple boundary interpolation problem for contracting matrix–valued functions in the unit circle*, Teoriya Funktsii, Funktsional'nyi Analiz i Ikh Prilozheniya, **51** (1989), 38–55. English transl. in: Journal of Soviet Mathematics, **52(6)** (1990), 3467–3481.
41. I. V. Kovalishina and V. P. Potapov, *Seven Papers Translated from the Russian*, American Mathematical Society Translations (2), **138**, Providence, R.I., 1988.
42. P. Lancaster and M. Tismenetsky, *The theory of matrices*, Academic Press, Orlando, 1985.
43. A. A. Nudelman, *On a new problem of moment problem type*, Sov. Math. Dokl., **18** (1977), 507–510.
44. Ch. Pommerenke, *Boundary Behaviour of Conformal Maps*, Springer-Verlag, Berlin, 1992.

45. M. Rosenblum and J. Rovnyak, *Hardy classes and operator theory*, Oxford University Press, 1985.
46. J. Rovnyak, *Characterization of spaces K(M)*, Unpublished manuscript, 1969.
47. S. Saitoh, *Theory of reproducing kernels and its applications*, **189** Longman, Essex, 1988.
48. D. Sarason, *Angular derivatives via Hilbert spaces*, Complex Variables, **10** (1988), 1–10.
49. D. Sarason, *Sub–Hardy Hilbert Spaces in the Unit Disk*, Wiley, New York, 1994.
50. D. Sarason, *Nevanlinna–Pick interpolation with boundary data*, Integral Equations Operator Theory, **30** (1998), 231–250.
51. J. H. Shapiro, *Composition operators and classical function theory*, Springer-Verlag, New York, 1993.

## Editorial Information

To be published in the *Memoirs*, a paper must be correct, new, nontrivial, and significant. Further, it must be well written and of interest to a substantial number of mathematicians. Piecemeal results, such as an inconclusive step toward an unproved major theorem or a minor variation on a known result, are in general not acceptable for publication. Papers appearing in *Memoirs* are generally at least 80 and not more than 200 published pages in length. Papers less than 80 or more than 200 published pages require the approval of the Managing Editor of the Transactions/Memoirs Editorial Board.

As of January 31, 2006, the backlog for this journal was approximately 14 volumes. This estimate is the result of dividing the number of manuscripts for this journal in the Providence office that have not yet gone to the printer on the above date by the average number of monographs per volume over the previous twelve months, reduced by the number of volumes published in four months (the time necessary for preparing a volume for the printer). (There are 6 volumes per year, each containing at least 4 numbers.)

A Consent to Publish and Copyright Agreement is required before a paper will be published in the *Memoirs*. After a paper is accepted for publication, the Providence office will send a Consent to Publish and Copyright Agreement to all authors of the paper. By submitting a paper to the *Memoirs*, authors certify that the results have not been submitted to nor are they under consideration for publication by another journal, conference proceedings, or similar publication.

## Information for Authors

*Memoirs* are printed from camera copy fully prepared by the author. This means that the finished book will look exactly like the copy submitted.

The paper must contain a *descriptive title* and an *abstract* that summarizes the article in language suitable for workers in the general field (algebra, analysis, etc.). The *descriptive title* should be short, but informative; useless or vague phrases such as "some remarks about" or "concerning" should be avoided. The *abstract* should be at least one complete sentence, and at most 300 words. Included with the footnotes to the paper should be the 2000 *Mathematics Subject Classification* representing the primary and secondary subjects of the article. The classifications are accessible from www.ams.org/msc/. The list of classifications is also available in print starting with the 1999 annual index of *Mathematical Reviews*. The Mathematics Subject Classification footnote may be followed by a list of *key words and phrases* describing the subject matter of the article and taken from it. Journal abbreviations used in bibliographies are listed in the latest *Mathematical Reviews* annual index. The series abbreviations are also accessible from www.ams.org/publications/. To help in preparing and verifying references, the AMS offers MR Lookup, a Reference Tool for Linking, at www.ams.org/mrlookup/. When the manuscript is submitted, authors should supply the editor with electronic addresses if available. These will be printed after the postal address at the end of the article.

**Electronically prepared manuscripts.** The AMS encourages electronically prepared manuscripts, with a strong preference for $\mathcal{A}\mathcal{M}\mathcal{S}$-LaTeX. To this end, the Society has prepared $\mathcal{A}\mathcal{M}\mathcal{S}$-LaTeX author packages for each AMS publication. Author packages include instructions for preparing electronic manuscripts, the *AMS Author Handbook*, samples, and a style file that generates the particular design specifications of that publication series. Though $\mathcal{A}\mathcal{M}\mathcal{S}$-LaTeX is the highly preferred format of TeX, author packages are also available in $\mathcal{A}\mathcal{M}\mathcal{S}$-TeX.

Authors may retrieve an author package from e-MATH starting from www.ams.org/tex/ or via FTP to ftp.ams.org (login as anonymous, enter username as password, and type cd pub/author-info). The *AMS Author Handbook* and the *Instruction Manual* are available in PDF format following the author packages link from www.ams.org/tex/. The author package can also be obtained free of charge by sending email to tech-support@ams.org (Internet) or from the Publication Division, American Mathematical Society, 201 Charles St., Providence, RI 02904-2294, USA. When requesting an author package, please specify $\mathcal{AMS}$-LaTeX or $\mathcal{AMS}$-TeX and the publication in which your paper will appear. Please be sure to include your complete mailing address.

**Sending electronic files.** After acceptance, the source file(s) should be sent to the Providence office (this includes any TeX source file, any graphics files, and the DVI or PostScript file).

Before sending the source file, be sure you have proofread your paper carefully. The files you send must be the EXACT files used to generate the proof copy that was accepted for publication. For all publications, authors are required to send a printed copy of their paper, which exactly matches the copy approved for publication, along with any graphics that will appear in the paper.

TeX files may be submitted by email, FTP, or on diskette. The DVI file(s) and PostScript files should be submitted only by FTP or on diskette unless they are encoded properly to submit through email. (DVI files are binary and PostScript files tend to be very large.)

Electronically prepared manuscripts can be sent via email to pub-submit@ams.org (Internet). The subject line of the message should include the publication code to identify it as a Memoir. TeX source files, DVI files, and PostScript files can be transferred over the Internet by FTP to the Internet node e-math.ams.org (130.44.1.100).

**Electronic graphics.** Comprehensive instructions on preparing graphics are available at www.ams.org/jourhtml/graphics.html. A few of the major requirements are given here.

Submit files for graphics as EPS (Encapsulated PostScript) files. This includes graphics originated via a graphics application as well as scanned photographs or other computer-generated images. If this is not possible, TIFF files are acceptable as long as they can be opened in Adobe Photoshop or Illustrator. No matter what method was used to produce the graphic, it is necessary to provide a paper copy to the AMS.

Authors using graphics packages for the creation of electronic art should also avoid the use of any lines thinner than 0.5 points in width. Many graphics packages allow the user to specify a "hairline" for a very thin line. Hairlines often look acceptable when proofed on a typical laser printer. However, when produced on a high-resolution laser imagesetter, hairlines become nearly invisible and will be lost entirely in the final printing process.

Screens should be set to values between 15% and 85%. Screens which fall outside of this range are too light or too dark to print correctly. Variations of screens within a graphic should be no less than 10%.

**Inquiries.** Any inquiries concerning a paper that has been accepted for publication should be sent directly to the Electronic Prepress Department, American Mathematical Society, 201 Charles St., Providence, RI 02904, USA.

# Editors

This journal is designed particularly for long research papers, normally at least 80 pages in length, and groups of cognate papers in pure and applied mathematics. Papers intended for publication in the *Memoirs* should be addressed to one of the following editors. In principle the Memoirs welcomes electronic submissions, and some of the editors, those whose names appear below with an asterisk (*), have indicated that they prefer them. However, editors reserve the right to request hard copies after papers have been submitted electronically. Authors are advised to make preliminary email inquiries to editors about whether they are likely to be able to handle submissions in a particular electronic form.

*Algebra to ALEXANDER KLESHCHEV, Department of Mathematics, University of Oregon, Eugene, OR 97403-1222; email: ams@noether.uoregon.edu

Algebra and its application to MINA TEICHER, Emmy Noether Research Institute for Mathematics, Bar-Ilan University, Ramat-Gan 52900, Israel; email: teicher@macs.biu.ac.il

Algebraic geometry to DAN ABRAMOVICH, Department of Mathematics, Brown University, Box 1917, Providence, RI 02912; email: amsedit@math.brown.edu

*Algebraic number theory to V. KUMAR MURTY, Department of Mathematics, University of Toronto, 100 St. George Street, Toronto, ON M5S 1A1, Canada; email: murty@math.toronto.edu

*Algebraic topology to ALEJANDRO ADEM, Department of Mathematics, University of British Columbia, Room 121, 1984 Mathematics Road, Vancouver, British Columbia, Canada V6T 1Z2; email: adem@math.ubc.ca

Combinatorics to JOHN R. STEMBRIDGE, Department of Mathematics, University of Michigan, Ann Arbor, Michigan 48109-1109; email: FRS@umich.edu

Complex analysis and harmonic analysis to ALEXANDER NAGEL, Department of Mathematics, University of Wisconsin, 480 Lincoln Drive, Madison, WI 53706-1313; email: nagel@math.wisc.edu

*Differential geometry and global analysis to LISA C. JEFFREY, Department of Mathematics, University of Toronto, 100 St. George St., Toronto, ON Canada M5S 3G3; email: jeffrey@math.toronto.edu

Dynamical systems and ergodic theory to AMIE WILKINSON, Department of Mathematics, Northwestern University, 2033 Sheridan Road, Evanston, IL 60208-2730; email: wilkinso@math.northwestern.edu

*Functional analysis and operator algebras to MARIUS DADARLAT, Department of Mathematics, Purdue University, 150 N. University St., West Lafayette, IN 47907-2067; email: mdd@math.purdue.edu

*Geometric analysis to TOBIAS COLDING, Courant Institute, New York University, 251 Mercer St., New York, NY 10012; email: traneditor@cims.nyu.edu

*Geometric analysis to MLADEN BESTVINA, Department of Mathematics, University of Utah, 155 South 1400 East, JWB 233, Salt Lake City, Utah 84112-0090; email: bestvina@math.utah.edu

Harmonic analysis, representation theory, and Lie theory to ROBERT J. STANTON, Department of Mathematics, The Ohio State University, 231 West 18th Avenue, Columbus, OH 43210-1174; email: stanton@math.ohio-state.edu

*Logic to STEFFEN LEMPP, Department of Mathematics, University of Wisconsin, 480 Lincoln Drive, Madison, Wisconsin 53706-1388; email: lempp@math.wisc.edu

*Ordinary differential equations, and applied mathematics to PETER W. BATES, Department of Mathematics, Michigan State University, East Lansing, MI 48824-1027; email: bates@math.msu.edu

*Partial differential equations to GUSTAVO PONCE, Department of Mathematics, South Hall, Room 6607, University of California, Santa Barbara, CA 93106; email: ponce@math.ucsb.edu

*Probability and statistics to KRZYSZTOF BURDZY, Department of Mathematics, University of Washington, Box 354350, Seattle, Washington 98195-4350; email: burdzy@math.washington.edu

*Real analysis and partial differential equations to DANIEL TATARU, Department of Mathematics, University of California, Berkeley, Berkeley, CA 94720; email: tataru@math.berkeley.edu

All other communications to the editors should be addressed to the Managing Editor, ROBERT GURALNICK, Department of Mathematics, University of Southern California, Los Angeles, CA 90089-1113; email: guralnic@math.usc.edu.

# Titles in This Series

856 **Vladimir Bolotnikov and Harry Dym,** On boundary interpolation for matrix valued Schur functions, 2006

855 **Yevgenia Kashina, Yorck Sommerhäuser, and Yongchang Zhu,** On higher Frobenius-Schur indicators, 2006

854 **Noam Greenberg,** The role of true finiteness in the admissible recursively enumerable degrees, 2006

853 **Joachim Krieger,** Stability of spherically symmetric wave maps, 2006

852 **Viorel Barbu, Irena Lasiecka, and Roberto Triggiani,** Tangential boundary stabilization of Navier-Stokes equations, 2006

851 **Jie Wu,** On maps from loop suspensions to loop spaces and the shuffle relations on the Cohen groups, 2006

850 **Siegfried Echterhoff, S. Kaliszewski, John Quigg, and Iain Raeburn,** A categorical approach to imprimitivity theorems for $C^*$-dynamical systems, 2006

849 **Katsuhiko Kuribayashi, Mamoru Mimura, and Tetsu Nishimoto,** Twisted tensor products related to the cohomology of the classifying spaces of loop groups, 2006

848 **Bob Oliver,** Equivalences of classifying spaces completed at the prime two, 2006

847 **Eric T. Sawyer and Richard L. Wheeden,** Hölder continuity of weak solutions to subelliptic equations with rough coefficients, 2006

846 **Victor Beresnevich, Detta Dickinson, and Sanju Velani,** Measure theoretic laws for lim–sup sets, 2006

845 **Ehud Friedgut, Vojtech Rödl, Andrzej Ruciński, and Prasad V. Tetali,** A Sharp threshold for random graphs with a monochromatic triangle in every edge coloring, 2006

844 **Amadeu Delshams, Rafael de la Llave, and Tere M. Seara,** A geometric mechanism for diffusion in Hamiltonian systems overcoming the large gap problem: Heuristics and rigorous verification on a model, 2006

843 **Denis V. Osin,** Relatively hyperbolic groups: Intrinsic geometry, algebraic properties, and algorithmic problems, 2006

842 **David P. Blecher and Vrej Zarikian,** The calculus of one-sided $M$-ideals and multipliers in operator spaces, 2006

841 **Enrique Artal Bartolo, Pierrette Cassou-Noguès, Ignacio Luengo, and Alejandro Melle Hernández,** Quasi-ordinary power series and their zeta functions, 2005

840 **Sławomir Kołodziej,** The complex Monge-Ampère equation and pluripotential theory, 2005

839 **Mihai Ciucu,** A random tiling model for two dimensional electrostatics, 2005

838 **V. Jurdjevic,** Integrable Hamiltonian systems on complex Lie groups, 2005

837 **Joseph A. Ball and Victor Vinnikov,** Lax-Phillips scattering and conservative linear systems: A Cuntz-algebra multidimensional setting, 2005

836 **H. G. Dales and A. T.-M. Lau,** The second duals of Beurling algebras, 2005

835 **Kiyoshi Igusa,** Higher complex torsion and the framing principle, 2005

834 **Kenichi Ohshika,** Kleinian groups which are limits of geometrically finite groups, 2005

833 **Greg Hjorth and Alexander S. Kechris,** Rigidity theorems for actions of product groups and countable Borel equivalence relations, 2005

832 **Lee Klingler and Lawrence S. Levy,** Representation type of commutative Noetherian rings III: Global wildness and tameness, 2005

831 **K. R. Goodearl and F. Wehrung,** The complete dimension theory of partially ordered systems with equivalence and orthogonality, 2005

830 **Jason Fulman, Peter M. Neumann, and Cheryl E. Praeger,** A generating function approach to the enumeration of matrices in classical groups over finite fields, 2005

829 **S. G. Bobkov and B. Zegarlinski,** Entropy bounds and isoperimetry, 2005

## TITLES IN THIS SERIES

828 **Joel Berman and Paweł M. Idziak,** Generative complexity in algebra, 2005
827 **Trevor A. Welsh,** Fermionic expressions for minimal model Virasoro characters, 2005
826 **Guy Métivier and Kevin Zumbrun,** Large viscous boundary layers for noncharacteristic nonlinear hyperbolic problems, 2005
825 **Yaozhong Hu,** Integral transformations and anticipative calculus for fractional Brownian motions, 2005
824 **Luen-Chau Li and Serge Parmentier,** On dynamical Poisson groupoids I, 2005
823 **Claus Mokler,** An analogue of a reductive algebraic monoid whose unit group is a Kac-Moody group, 2005
822 **Stefano Pigola, Marco Rigoli, and Alberto G. Setti,** Maximum principles on Riemannian manifolds and applications, 2005
821 **Nicole Bopp and Hubert Rubenthaler,** Local zeta functions attached to the minimal spherical series for a class of symmetric spaces, 2005
820 **Vadim A. Kaimanovich and Mikhail Lyubich,** Conformal and harmonic measures on laminations associated with rational maps, 2005
819 **F. Andreatta and E. Z. Goren,** Hilbert modular forms: Mod $p$ and $p$-adic aspects, 2005
818 **Tom De Medts,** An algebraic structure for Moufang quadrangles, 2005
817 **Javier Fernández de Bobadilla,** Moduli spaces of polynomials in two variables, 2005
816 **Francis Clarke,** Necessary conditions in dynamic optimization, 2005
815 **Martin Bendersky and Donald M. Davis,** $V_1$-periodic homotopy groups of $SO(n)$, 2004
814 **Johannes Huebschmann,** Kähler spaces, nilpotent orbits, and singular reduction, 2004
813 **Jeff Groah and Blake Temple,** Shock-wave solutions of the Einstein equations with perfect fluid sources: Existence and consistency by a locally inertial Glimm scheme, 2004
812 **Richard D. Canary and Darryl McCullough,** Homotopy equivalences of 3-manifolds and deformation theory of Kleinian groups, 2004
811 **Ottmar Loos and Erhard Neher,** Locally finite root systems, 2004
810 **W. N. Everitt and L. Markus,** Infinite dimensional complex symplectic spaces, 2004
809 **J. T. Cox, D. A. Dawson, and A. Greven,** Mutually catalytic super branching random walks: Large finite systems and renormalization analysis, 2004
808 **Hagen Meltzer,** Exceptional vector bundles, tilting sheaves and tilting complexes for weighted projective lines, 2004
807 **Carlos A. Cabrelli, Christopher Heil, and Ursula M. Molter,** Self-similarity and multiwavelets in higher dimensions, 2004
806 **Spiros A. Argyros and Andreas Tolias,** Methods in the theory of hereditarily indecomposable Banach spaces, 2004
805 **Philip L. Bowers and Kenneth Stephenson,** Uniformizing dessins and Belyĭ maps via circle packing, 2004
804 **A. Yu Ol'shanskii and M. V. Sapir,** The conjugacy problem and Higman embeddings, 2004
803 **Michael Field and Matthew Nicol,** Ergodic theory of equivariant diffeomorphisms: Markov partitions and stable ergodicity, 2004
802 **Martin W. Liebeck and Gary M. Seitz,** The maximal subgroups of positive dimension in exceptional algebraic groups, 2004
801 **Fabio Ancona and Andrea Marson,** Well-posedness for general $2 \times 2$ systems of conservation law, 2004

For a complete list of titles in this series, visit the
AMS Bookstore at **www.ams.org/bookstore/**.